Topics in
Current Physics 28

Topics in Current Physics Founded by Helmut K. V. Lotsch

1 **Beam-Foil Spectroscopy**
Editor: S. Bashkin

2 **Modern Three-Hadron Physics**
Editor: A. W. Thomas

3 **Dynamics of Solids and Liquids by Neutron Scattering**
Editors: S. W. Lovesey and T. Springer

4 **Electron Spectroscopy for Surface Analysis**
Editor: H. Ibach

5 **Structure and Collisions of Ions and Atoms**
Editor: I. A. Sellin

6 **Neutron Diffraction**
Editor: H. Dachs

7 **Monte Carlo Methods** in Statistical Physics
Editor: K. Binder

8 **Ocean Acoustics**
Editor: J. A. DeSanto

9 **Inverse Source Problems** in Optics
Editor: H. P. Baltes

10 **Synchrotron Radiation**
Techniques and Applications
Editor: C. Kunz

11 **Raman Spectroscopy** of Gases and Liquids
Editor: A. Weber

12 **Positrons in Solids**
Editor: P. Hautojärvi

13 **Computer Processing of Electron Microscope Images**
Editor: P. W. Hawkes

14 **Excitons** Editor: K. Cho

15 Physics of **Superionic Conductors**
Editor: M. B. Salamon

16 **Aerosol Microphysics I**
Particle Interactions
Editor: W. H. Marlow

17 **Solitons**
Editors: R. Bullough, P. Caudrey

18 **Properties of Magnetic Electron Lenses**
Editor: P. W. Hawkes

19 Theory of **Chemisorption**
Editor: J. R. Smith

20 **Inverse Scattering Problems** in Optics
Editor: H. P. Baltes

21 **Coherent Nonlinear Optics**
Recent Advances
Editors: M. S. Feld and V. S. Letokhov

22 **Electromagnetic Theory of Gratings**
Editor: R. Petit

23 **Structural Phase Transitions I**
Editors: K. A. Müller and H. Thomas

24 **Amorphous Solids**
Low-Temperature Properties
Editor: W. A. Phillips

25 **Mössbauer Spectroscopy II**
The Exotic Side of the Method
Editor: U. Gonser

26 **Crystal Cohesion and Conformational Energies** Editor: R. M. Metzger

27 **Dissipative Systems in Quantum Optics**
Resonance Fluorescence, Optical Bistability, Superfluorescence
Editor: R. Bonifacio

28 **The Stratospheric Aerosol Layer**
Editor: R. C. Whitten

29 **Aerosol Microphysics II**
Chemical Physics of Microparticles
Editor: W. H. Marlow

30 **Real Space Renormalization**
Editors: J. M. van Leeuwen and T. W. Burkhardt

31 **Hyperfine Interactions of Radioactive Nuclei**
Editor: J. Christiansen

32 **Superconductivity in Ternary Compounds I**
Structural, Electronic, and Lattice Properties
Editors: B. Maple and O. Fischer

The Stratospheric Aerosol Layer

Edited by R.C. Whitten

With Contributions by
A. W. Castleman   W. P. Chu   N. H. Farlow
P. Hamill   E. C. Y. Inn   R. G. Keesee
M. P. McCormick   J. B. Pollack   P. B. Russell
O. B. Toon   R. P. Turco   R. C. Whitten

With 62 Figures

Springer-Verlag Berlin Heidelberg New York 1982

Professor Dr. Robert C. Whitten
Space Science Division, NASA-Ames Research Center
Moffett Field, CA 94035, USA

ISBN 3-540-11229-4 Springer-Verlag Berlin Heidelberg New York
ISBN 0-387-11229-4 Springer-Verlag New York Heidelberg Berlin

Library of Congress Cataloging in Publication Data. Main entry under title: The Stratospheric sulfate aerosol layer. (Topics in current physics ; 28) Includes bibliographical references and index. 1. Aerosols—Measurement. 2. Sulfates—Measurement. 3. Stratosphere—Measurement. 4. Atmospheric chemistry. I. Whitten, R. C. (Robert Craig), 1926— . II. Castleman, A. W. (Albert Welford), 1936- . III. Series. QC882.S77   551.5'142   81-21208   AACR2

This work is subject to copyright. All rights are reserved, whether the whole or part of the material is concerned, specifically those of translation, reprinting, reuse of illustrations, broadcasting, reproduction by photocopying machine or similar means, and storage in data banks. Under § 54 of the German Copyright Law where copies are made for other than private use, a fee is payable to "Verwertungsgesellschaft Wort", Munich.

© by Springer-Verlag Berlin Heidelberg 1982
Printed in Germany

The use of registered names, trademarks, etc. in this publication does not imply, even in the absence of a specific statement, that such names are exempt from the relevant protective laws and regulations and therefore free for general use.

Offset printing and bookbinding: Konrad Triltsch, Graphischer Betrieb, Würzburg.
2153/3130-543210

Preface

It is now a quarter of a century since Junge and his coworkers recovered the first sample from the sulfate aerosol layer in the stratosphere. Since that time vast strides have been made in determining its physical properties and morphology. These investigations have been performed with instruments on board aircraft and balloon platforms as in the early days, with ground-based lidar (optical radar), and most recently with satellite-borne optical instruments. It will become evident in Chapter 2 that in situ measurements by aircraft and balloon sensors complement rather than duplicate the remote techniques (lidar and satellite). Hence future programs will probably continue to utilize direct as well as indirect experimental techniques.

Concurrently, with the observations of the gross properties of the aerosol layer, laboratory and theoretical studies have sought to elucidate the chemical and microphysical processes which influence the formation and growth of the aerosol particles. The laboratory investigations have included studies of gas phase chemistry, and particle nucleation and growth mechanisms. Theoretical studies have revolved mainly around a series of models developed by atmospheric scientists. The earliest of these models was constructed by Junge and his colleagues. With the advent of third- and fourth-generation computers, the capacity to solve the quite complex continuity equations which govern particle formation, growth, and removal has advanced to the point where most of the particle properties can be simulated with reasonable confidence.

The potential importance of a light-scattering particle layer in the stratosphere is obvious: perturbations of the layer, especially due to violent volcanic eruptions, can lead to substantial, if short-term, alterations in the mean surface temperature of the Earth. In fact, the observational evidence for such correlations between volcanic activity and climate fluctuations is compelling. Only recently has man's influence on stratospheric aerosols, and thus the climate, come under close scrutiny and evaluation.

The time for a general review of the chemistry and physics of the stratospheric aerosol layer therefore appears to be opportune. The present book attempts to fill this need with a coordinated series of chapters by a group of investigators who are leaders in the field. An introductory chapter, which sets the general tone of the volume and gives synopses of current knowledge of the layer and the associated microphysical processes, is followed by four chapters which go into great detail on field measurements, associated chemical processes, theoretical models, and climate effects of aerosols. It is hoped that this treatise will serve as a useful reference for those who are engaged or interested in studying stratospheric aerosols.

Cupertino, California   *Robert C. Whitten*
November 1981

Contents

1. Introduction
   By R.C. Whitten and P. Hamill (With 1 Figure) 1
   1.1 General Discussion 1
       1.1.1 Historical 1
       1.1.2 Particle Measurements 1
       1.1.3 Measurements of Precursor Gases 4
       1.1.4 Laboratory Measurements 5
       1.1.5 Models of the Aerosol Layer 6
   1.2 Influence on Climate 6
   1.3 Microphysical Processes 7
       1.3.1 Nucleation 7
       1.3.2 Condensation and Evaporation 9
       1.3.3 Coagulation 10
       1.3.4 Sedimentation 11
       1.3.5 Transport 12
   References 13

2. Observations. By E.C.Y. Inn, N.H. Farlow, P.B. Russell, M.P. McCormick, and W.P. Chu (With 27 Figures) 15
   2.1 Measurements of Precursor Gases in the Stratosphere 15
       2.1.1 In Situ Mass Spectrometer Method 15
       2.1.2 In Situ Filter Collection 16
       2.1.3 In Situ Cryogenic Collection 17
       2.1.4 Remote Sensing Methods 21
       2.1.5 Summary of Stratospheric Observations of Sulfur Gases 21
   2.2 Aircraft and Balloon Measurement of Aerosol Properties 23
       2.2.1 Aerosol Constituents 23
       2.2.2 Aerosol Properties 23
       2.2.3 Detection Methods 24
       2.2.4 Analytical Techniques 25
       2.2.5 Statistical Treatments 26
       2.2.6 Aitken Nuclei 27

		2.2.7 Larger Acid Droplets	27
		2.2.8 Granular Inclusions	30
		2.2.9 Cosmic Dust	30
		2.2.10 Volcanic Ash	31
		2.2.11 Variations	31
		2.2.12 Discussion	33
	2.3	Lidar Measurements	34
		2.3.1 The Lidar Technique	35
		2.3.2 Survey of Results	37
		2.3.3 Comparisons of Lidar and Other Results	43
		2.3.4 Additional Applications	46
	2.4	Satellite Observations	47
		2.4.1 Present Spacecraft Experiments	49
		2.4.2 Orbital Considerations	52
		2.4.3 Results Obtained with SAM II	54
		2.4.4 Results Obtained with SAGE	60
		2.4.5 Ground-Truth Comparisons	62
		2.4.6 Applications	64
	References		64

3. The Chemical Kinetics of Aerosol Formation
By R.G. Keesee and A.W. Castleman, Jr. (With 3 Figures) 69

3.1	Chemical Origin	69
3.2	Nucleation	74
3.3	Nucleation Mechanisms in the Stratosphere	78
	3.3.1 Binary Nucleation	79
	3.3.2 Ternary Nucleation	80
	3.3.3 Binary Heterogeneous Nucleation	80
	3.3.4 Heteromolecular Nucleation	81
3.4	Growth and Heterogeneous Reactions	84
3.5	Conclusions	90
References		90

4. Models of Stratospheric Aerosols and Dust
By R.P. Turco (With 16 Figures) 93

4.1	Overview	93
4.2	The Generalized Aerosol Continuity Equation	95
4.3	Aerosol Models	100
	4.3.1 The Simulated Distributions of Aerosol Precursor Gases	102
	4.3.2 Aerosol Nucleation Models	105
	4.3.3 Calculated Properties of the Aerosols	106

	4.4 Models of Upper Atmospheric Dust	111
	4.5 Anthropogenic Perturbations of the Aerosols	113
	4.6 Conclusions	117
	References	117

5. Stratospheric Aerosols and Climate
By O.B. Toon and J.B. Pollack (With 15 Figures) 121

5.1	Background	121
5.2	Statistical Relations Between Volcanic Explosions and Climatic Changes	123
	5.2.1 Changes Observed After Single Eruptions	123
	5.2.2 Changes Observed During Epochs of Volcanic Activity	127
5.3	Theoretical Relationships Between Volcanic Explosions and Climate	133
	5.3.1 Radiative Properties of the Aerosols	133
	5.3.2 Sensitivity Studies of the Effects of Aerosols on Climate	140
	5.3.3 Theoretical Studies of the Effect of Volcanoes on Climate	143
5.4	Studies of Anthropogenic Alterations of the Stratospheric Aerosol Layer	144
5.5	Summary	145
	References	146

Subject Index 149

List of Contributors

Castleman, A.W., Jr.
   University of Colorado, Boulder, CO 80309, USA

Chu, William P.
   Atmospheric Environmental Sciences Division, NASA-Langley Research Center, Hampton, VA 23665, USA

Farlow, Neil H.
   Space Science Division, NASA-Ames Research Center, Moffett Field, CA 94035, USA

Hamill Patrick
   Department of Physics, San Jose State University, San Jose, CA 95192, USA

Inn, Edward C.Y.
   Space Science Division, NASA-Ames Research Center, Moffett Field, CA 94035, USA

Keesee, Robert G.
   Space Science Division, NASA-Ames Research Center, Moffett Field, CA 94035, USA

McCormick, M. Patrick
   Atmospheric Environmental Sciences Division, NASA-Langley Research Center, Hampton, VA 23665, USA

Pollack, James B.
   Space Science Division, NASA-Ames Research Center, Moffett Field, CA 94035, USA

Russell, Philip B.
   SRI International, Menlo Park, CA 94025, USA

Toon, Owen B.
   Space Science Division, NASA-Ames Research Center, Moffett Field, CA 94035, USA

Turco, Richard P.
   R and D Associates, Marina del Rey, CA 90291, USA

Whitten, Robert C.
   Space Science Division, NASA-Ames Research Center, Moffett Field, CA 94035 CA 94035, USA

1. Introduction

R. C. Whitten and P. Hamill

With 1 Figure

1.1 General Discussion

1.1.1 Historical

The existence of a layer of particles in the stratosphere was suggested many years ago by GRUNER and KLEINERT [1.1], who based their proposal on twilight observations of the purple light. The direct measurement of the properties of the layer had to await the development of high-altitude research aircraft and balloons, which could fly routinely into the lower stratosphere. JUNGE and co-workers [1.2] discovered the layer (in the sense of actually collecting particles) while carrying out high-altitude radioactivity measurements. Since then, a vast store of information has been collected by various techniques (to be discussed later), which establishes the composition, size distributions, and temporal variations of the layer. The physical origin of the aerosols is nearly as well understood, being due mainly to the condensation of sulfates produced by the solar photodissociation of carbonyl sulfide (OCS) by ultraviolet light [1.3] and by the injection of sulfur dioxide into the stratosphere during volcanic eruptions [1.4]. While the earliest methods of direct aerosol sampling by collection on impaction plates, wires, or filters are still in use [1.5-8], optical techniques have been introduced since the work of JUNGE. These include optical particle counters [1.9], optical radar (frequently called *lidar*) [1.10-13], and measurement of the attenuation of solar radiation during satellite occultation [1.14].

1.1.2 Particle Measurements

The technique employed in the pioneering work of JUNGE et al. [1.2] used cascade impactors, gummed glass slides as collector plates with balloons as the instrumental platforms. Interpretation of the balloon sampling data was hampered by the fact that the sample was obtained with a small volume of air taken at a specific time and location. It was fortunate, therefore, that about two years later high-altitude research aircraft (U-2) became available to the program. In their newer instrument designed for aircraft mounting, JUNGE and MANSON [1.15] used stainless steel particle collector plates as impaction surfaces. The plates were contained within flow tubes designed to channel ambient air over the collection surfaces. The system, which was characterized by close tolerances, was otherwise completely enclosed so that it was isolated from the environment when the seals were in place. Size distributions were determined for altitudes ranging from 12 to 21 km by measuring particle sizes from electron micrographs and performing the appropriate statistical analyses. Absolute particle number densities were determined from particle counts and aerodynamic parameters (collection efficiency, air flow speed, and exposure time). Subsequent chemical analysis revealed that the particles contained sulfur as a major constituent.

BIGG et al. [1.16] also used an impaction technique, but mounted their instruments on balloons rather than aircraft. As a result, they were able to extend the region of investigation upward to 37 km. Their collection technique was basically similar to that of JUNGE and coworkers, with the difference being mainly in the means of producing air flow. JUNGE and MANSON [1.15] used the ram pressure induced by aircraft motion, whereas BIGG et al. used a pump to create the air flow.

LAZRUS et al. [1.17] utilized cellulose filters impregnated with dibutoxy ethylphthalate for collection of sulfates. After each flight at 18-19 km altitude on a RB57-F aircraft, the filters were ultrasonically rinsed in distilled water containing a nonionic detergent. The rinse water was then analyzed for sulfate and ammonium ions. They found that only a small fraction of the ions were ammonium, which they attributed to contamination by tropospheric air.

FARLOW and co-workers [1.6,7] adopted a system of carbon-coated palladium wires mounted on a high-altitude aircraft (U-2) to measure aerosol properties at 18-21 km altitude. After each flight the wires were taken to the laboratory and examined with a scanning electron microscope in order to obtain size distributions and absolute particle concentrations. Earlier analysis of the composition [1.18] revealed that they contained ammonium sulfate. However, after modifying the collectors so that they were not exposed to the atmosphere after collection, it was found that the ammonium crystals were not present. Apparently they were formed in the lower atmosphere *after* collection [1.19], which also explains the result of LAZRUS et al. [1.17].

The preceding techniques depend upon direct collection of particles which are then returned to the laboratory for analysis. Because of the possibility of evaporation of some of the collected material while in transit to the laboratory, ROSEN [1.9] developed an in situ optical method for observing the particles. He used a photomultiplier counter equipped with the necessary optics and a pulse height analyzer to actually count particles as they flowed through the system. If the particles are of sufficient size, the photomultiplier will respond with an electrical pulse, the pulse height being a function of the particle size and its index of refraction. From these data, a crude approximation to the particle size spectrum can be determined. Of course, the spectrum is not uniquely determined because its shape and the particle composition (index of refraction) must be assumed. Nevertheless, the system is simple, reliable, and easily mounted on balloons.

In addition to in situ sampling of the aerosol layer, remote sensing by ground-, aircraft-, and satellite-based systems is now in progress. The *lidar* systems previously mentioned employ pulsed laser beams which are projected vertically into the atmosphere. The intensity of the light which is backscattered by atmospheric gases and particles is measured as a function of altitude by means of a telescope and photomultiplier system. Altitude is determined by the time delay between pulse emission and receipt of the scattered signal. Although the laser beam is initially highly concentrated, it does spread with distance so that the signal-to-noise ratio for measurements of stratospheric aerosol layers is rather low; there are, however, techniques for improving the ratio. The data obtained by lidar are used to deduce aerosol properties such as turbidity and optical depth with the aid of an optical model.

The last technique reviewed here is based on satellite observations of the occultation of the sun by aerosols in the Earth's atmosphere. The first such experiment was carried out during the Apollo-Soyuz program in 1975 [1.20]. The instrument used was a simple photometer which the astronauts pointed at the sun during satellite sunrises and sunsets. Extinction coefficients were obtained by numerically inverting the sunlight intensity variation that was recorded in a band centered at 0.85 μm. The more recent SAM II and SAGE (Satellite Aerosol and Gas Experiment) programs use a similar technique (at a slightly different wavelength); data acquisition is necessarily automated because the spacecraft are unmanned [1.14].

It is interesting to note that several aerosol investigators have accumulated enough data at various times and places to define meridional and temporal variations of the layer as well as the local random variability. For example, LAZRUS and GANDRUD [1.8] used filters in a series of measurements extending from 51° South to 75° North latitude at various seasons in 1971 to 1973 (a volcanically quiet period). They found that the sulfate was rather evenly distributed between the two hemispheres and that higher sulfate concentrations occurred at higher altitudes in the tropics than near the poles. They also concluded from their observations that stratospheric sulfate is apparently carried downward into the troposphere in the vicinity of tropospheric folds.

HOFMANN et al. [1.21] employed an optical particle counter in a series of balloon flights over Laramie, Wyoming, in 1972 and 1973 (a quiescent period). They found the summer aerosol layer to be highly stable, with both the upper troposphere and lower stratosphere relatively free of particles. In winter, the tropopause height decreased and the layer itself became less uniform; moreover, the aerosols in the lower stratosphere appeared to increase in concentration. The layer was found to be somewhat variable in spring with evidence for a "complicated stratospheric-tropospheric exchange" process. HOFMANN et al. also concluded that particle removal from the stratosphere occurs principally in spring and summer. They also reported latitudinal variations in the aerosols [1.22] at various seasons (1972 to early 1974) in the Northern Hemisphere. Like LAZRUS and GANDRUD, they found that the peak concentration occurred at higher altitudes in the tropics than near the poles. Moreover, they found that the peak concentrations in the tropics were larger than the poleward peak concentrations. They tentatively interpreted the latter finding as being consistent with the hypothesis of an equatorial source of aerosols for the stratosphere.

FARLOW et al. [1.7] used their impact collectors discussed earlier to measure the latitudinal variations of the properties of the aerosol layer. Like the other investigators, they found the layer peak to be higher at low latitudes than at high latitudes. They also concluded that the particles are formed mainly at tropical latitudes. In more recent work, FARLOW and coworkers [1.23] obtained evidence that the particles grow and mature as they rise through the stratosphere. This observation makes it more difficult to ascribe a definite latitudinal zone of origin.

RUSSELL et al. ([1.24], also [1.13]) made a series of lidar observations of the aerosol layer following the Fuego volcanic eruption in October 1974. They began their measurements in February 1975, the time of maximum influence of the volcanic injection. The observed "e-folding" time of the vertically integrated particulate backscattering and the peak ratio of particulate to gaseous backscattering were found to be 8 and 11 months, respectively. The short lifetimes were evidently the result of the low altitude of the volcanically generated particles and their larger mean size compared to the ambient aerosol particles.

The preceding discussion of the results of several measurement programs is not intended to be exhaustive, but merely to give the flavor of the type of investigations required to establish the properties of the aerosol layer under both quiescent and volcanically perturbed conditions. A complete treatment is deferred to Chap.2.

1.1.3 Measurements of Precursor Gases

As we mentioned earlier, the stratospheric aerosol particles are formed from precursors consisting of sulfur-bearing gases, of which sulfur dioxide and carbonyl sulfide are the most prominent. Sulfur dioxide, an important component of volcanic gases, is often injected directly into the stratosphere by volcanic eruptions. However, during long periods of little or no volcanic activity, carbonyl sulfide (OCS) is apparently the main sulfur precursor [1.3]. Suggestions [1.25-27] have been made that carbon disulfide (CS_2) is an important precursor compound, being converted to OCS by reaction with OH in the troposphere. However, laboratory measurements have demonstrated that the reaction of OH with CS_2 is too slow to contribute to the stratospheric layer [1.28,29]. In any event, TURCO et al. [1.30] have shown that the ambient level of atmospheric OCS is sufficient to account for nearly all of the properties of the quiescent aerosol layer.

The first measurements of high-altitude SO_2 mixing ratios were made by JAESCHKE and co-workers [1.31], who extended their techniques (collection by means of a wet chemical filter with subsequent chemiluminescent analysis) to heights just above the tropopause. More recent efforts using dry filters and chemiluminescent analysis [1.32] reached to about 15 km, yielding SO_2 mixing fractions in the range 0.01 to 0.1 ppbv (parts per billion by volume) with a mean value of about 0.05 ppbv above 6-km altitude. INN et al. [1.33], who trapped the gases cryogenically and analyzed the samples by means of gas chromotography, were able to extend the observations up to 21 km, the maximum flight altitude of the U-2 aircraft. They obtained results similar to those reported by GEORGII and MEIXNER [1.32], namely 0.036 to 0.051 ppbv.

INN et al. [1.33,34] have also measured the stratospheric abundance of OCS, finding about 0.3 to 0.5 ppbv at altitudes (~ 15 km) just above the tropopause, but decreasing rapidly with height above 15 km. These values are consistent with tropospheric measurements made during the "Gametag" program [1.35]. INN et al. [1.33] also looked for stratospheric CS_2 but were not able to definitely identify it. A small feature in their data may have indicated the presence of CS_2; if so, its abundance would have been about 0.001 ppbv, a value much too small to make it of any direct significance to the stratosphere.

Discussion of the measurement of precursor gas mixing fractions is given in much greater detail in Chap.2.

1.1.4 Laboratory Measurements

In order to understand the formation of aerosols in the stratosphere, it is necessary to know the details of the chemical processes which affect the layer. The oxidation of OCS to SO_2 is rather well known [1.3,36], but the further oxidation to sulfuric acid is not. It is generally thought that OH is an oxidizing agent for part of the chain leading to HSO_x radicals. However, the reactions of each of these radicals with each other and with other species have not been established. Eventually, the HSO_x molecules can either form SO_3, which reacts with water to form H_2SO_4 vapor, or condense on preexisting aerosols and decompose in solution to form aqueous H_2SO_4.

Nucleation is a chemical process which initiates sulfuric acid droplet formation from the gas phase. The mode of formation, whether by heterogeneous nucleation on some very small dust particles (condensation nuclei) or by some other mechanism such as radical clustering, is not yet definitely known. Since nucleation is a very complex subject which is difficult to describe adequately in the short space available here, we defer its discussion to Sect.1.3 and, in much greater detail, to Chap.3 where all aspects of the chemistry of the aerosol layer and pertinent laboratory measurements are discussed.

1.1.5 Models of the Aerosol Layer

Over the past twenty years a variety of models of the stratospheric sulfate aerosol layer have been constructed. JUNGE and co-workers [1.2] were limited by computational capabilities of the day to a very simple model employing essentially analytic methods to treat vertical eddy transport, sedimentation, and coagulation. Despite neglect of growth and evaporation, their analysis — with the aid of the model — revealed many of the important sources and processes of stratospheric aerosols and helped to explain several features of their experimental data, particularly the vertical distribution of the "large" particles (i.e., particles with radii between 0.15 and 2 μm). Beginning in the mid 1970's, more sophisticated models employing the computational capabilities of third-generation computers were developed by BURGMEIER and BLIFFORD [1.37], KRITZ [1.38,39], ROSEN et al. [1.40], and TURCO et al. [1.36]. The last, which is discussed in great detail in Chap.4, includes all of the processes thought to be important to the formation of the layer (i.e., nucleation, vertical eddy diffusion, sedimentation, growth and evaporation, and coagulation). WHITTEN et al. [1.41] have reviewed all four models.

1.2 Influence on Climate

The stratospheric aerosol absorbs and scatters light from the sun. Assuming that the particles are spherical, Mie-scattering computational techniques can be applied to this problem. The particles also absorb and scatter infrared (thermal) radiation from the Earth. Accordingly, changes in their abundance cause a temporary imbalance between the solar energy absorbed and the thermal radiation emitted to space. The energy balance is restored by a corresponding change in the globally averaged surface temperature.

POLLACK et al. [1.42-44] performed the appropriate radiative transfer calculations required to assess the effect on climate of volcanically produced particles and aerosols likely to be formed by the oxidation of the sulfur dioxide contained in supersonic transport exhaust. More recently, the model of TURCO et al. [1.36] was used to make improved computations of the climatic effects of sulfur dioxide and soot emitted by supersonic transports [1.45] and of increased anthropogenic production of carbonyl sulfide [1.30,41]. Chapter 5 treats the climatic influence of stratospheric particles in considerable detail.

1.3 Microphysical Processes

In this section we briefly describe the microphysical processes affecting the formation of the stratospheric aerosol layer. The mechanisms considered are nucleation, condensation and evaporation, coagulation, sedimentation, and transport. When all of these mechanisms are incorporated into a numerical model, the general characteristics of the stratospheric aerosol layer can be simulated quite well (Chap.4). Processes which have not yet been incorporated into theoretical simulations of the aerosol layer are horizontal transport processes which require the framework of two- or three-dimensional models. Indeed, the Hadley cell circulation and tropopause folding probably exert significant influences on the layer but have not yet been theoretically analyzed in terms of their effects on the stratospheric aerosol. It is anticipated that when more observational data have been collected (Chap.2), these dynamical processes will be included in the numerical model simulations which are such a valuable tool for determining the impact of perturbations on the layer. Aerosol microphysical processes are discussed in great detail in the recent work edited by MARLOW [1.46].

1.3.1 Nucleation

The formation of stratospheric aerosol particles requires some sort of gas-to-particle conversion in the stratosphere. Early attempts to describe the stratospheric aerosol layer as resulting from the upward transport of tropospheric particles which then become trapped in the stratosphere and coagulate to form large particles were not particularly successful [1.47]. When investigators began searching for a mechanism to explain the formation of the stratospheric particles, they found that the most probable explanation lay in the nucleation of the binary system of sulfuric acid and water. This formation mechanism is highly favored thermodynamically because sulfuric acid in solution has an extremely low vapor pressure; consequently, the nucleation of sulfuric acid droplets occurs even when the amount of gas phase sulfuric acid present is quite small.

There are three main nucleation processes by which the stratospheric particles can be formed. These are homogeneous nucleation, heterogeneous nucleation onto preexisting solid surfaces, and nucleation onto ions. Theoretical studies of the homogeneous nucleation rate for sulfuric acid solution droplets under stratospheric conditions were carried out by HIDY et al. [1.48] and by HAMILL et al. [1.49,50]. They reached very different conclusions. HIDY and co-workers found that homogeneous nucleation could be significant in the stratosphere while HAMILL et al. concluded that this process led to the generation of a negligible number of new particles. The variance in these results was due to the fact that the two groups had used different estimates for the amount of gas phase sulfuric acid in the stratosphere. The nucleation rate of sulfuric acid droplets is an extremely sensitive function of the environmental concentration of gas phase H_2SO_4, varying up to twenty orders of magnitude for a one-order-of-magnitude change in the acid concentration [1.49]. Thus, it is essential that any estimates of nucleation rates be based on the best available information on the gas phase concentration of sufuric acid. Unfortunately, there are no good measurements of this parameter in the stratosphere, but recent work by ARNOLD and his coworkers [1.51] may lead to credible sulfuric acid concentration profiles.

The nucleation of binary H_2SO_4-H_2O solutions onto stratospheric ions has often been mentioned as a possible source of the stratospheric aerosol particles, but as was pointed out by MOHNEN [1.52], the ion lifetime in the stratosphere is limited by recombination. Due to this effect, the ions do not last long enough to collect the sulfuric acid and water molecules necessary to act as nucleation embryos.

Stratospheric sulfate droplets may also be formed by binary system heterogeneous nucleation onto solid surfaces of particles which are transported into the stratosphere by upward mixing from the troposphere [1.36], or onto particles which are deposited in the stratosphere by meteoritic ablation [1.9,53]. Another possible formation mechanism for stratospheric aerosol particles was suggested by FRIEND et al. [1.54] who proposed that the particles result from the clustering of HSO_3^- and HSO_5^--hydrated radicals which are formed by the photooxidation of SO_2 in air.

HAMILL et al. [1.50] have incorporated these processes into a fully interactive one-dimensional aerosol model. The result of the computations, presented in Fig.1.1, is quite interesting. In the figure, nucleation rates are given for homogeneous, ion, and heterogeneous nucleation as well as for a simple parameterized nucleation scheme which was not based on the physics of the nucleation mechanism. Note that the scale of the nucleation rate spans sixty orders of magnitude. Clearly, the different nucleation mechanisms are likely to cause vastly different rates of particle production. In the stratosphere, there is no significant rate of particle production by homogeneous nucleation, and ion nucleation is completely negligible.

Strangely enough, the use of the nucleation calculations in a fully interactive model does not yield results which are distinguishable from those obtained using the parameterized nucleation scheme. That is, it would appear that in simulating an aerosol layer with a numerical model, it makes little difference which particular form of nucleation simulation is used, and indeed, the number of new particles formed per unit time is also unimportant. In fact, the number of new particles formed can be orders of magnitude different and still yield the same number of large particles. Conversely, this means that studying the gross characteristics of an aerosol gives little or no information regarding the nucleation mechanism responsible for the formation of the particles. The statements above are true only because present-day instrumentation does not allow the determination of aerosol characteristics which are affected by the nucleation mechanism responsible for particle formation. This would require measurements of the size distribution of particles smaller than 0.01-μm radius and the total number of particles larger than 0.01-μm radius.

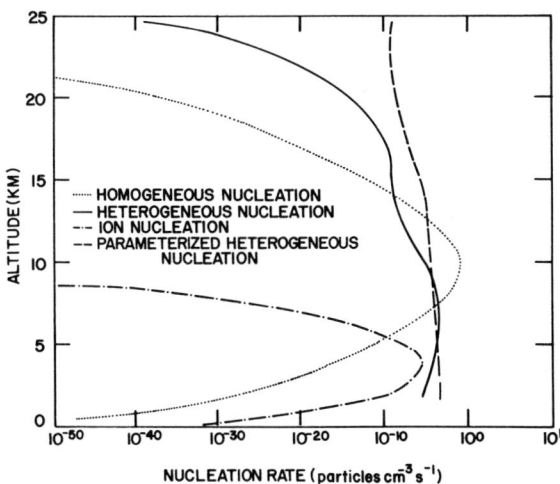

Fig.1.1. Nucleation rates for homogeneous nucleation, ion nucleation, and heterogeneous nucleation as a function of altitude as determined using a fully interactive one-dimensional aerosol model. Also shown for comparison purposes is a parameterized nucleation rate profile used in previous modeling work

In conclusion, it would appear from the results presented in Fig.1.1 that the principal particle formation mechanism in the stratosphere is the nucleation of sulfuric acid and water solutions onto preexisting solid particles, including the ion-ion recombinations postulated by ARNOLD [1.55]. The radical agglomeration mechanisms suggested by FRIEND et al. [1.54] can be neither proved nor disproved by model calculations and must therefore be considered a possible alternative mode of particle formation.

1.3.2 Condensation and Evaporation

Once a particle has been formed, it will either grow (or evaporate), depending upon local thermodynamic conditions. The mechanism for growth by this process is called *heteromolecular condensation*, for it requires the simultaneous gain of two different types of molecules.

Consider a sulfuric acid-water solution droplet suspended in air. It is subjected to a continuous bombardment by gas phase molecules, including water and sulfuric acid molecules. Under stratospheric conditions the droplet will be struck by millions of water molecules for every sulfuric acid molecule incident upon it. It is reasonable to assume that the droplet will tend towards an equilibrium state with respect to water such that the number of water molecules evaporating from the droplet per unit time is equal to the number of water molecules condensing on to the droplet per unit time. Thermodynamically, this means that the vapor pressure of water in the droplet is equal to the partial pressure of water in the surrounding air. In the stratosphere this equilibrium condition leads to droplets which are about 70% acid by weight, in good agreement with observations [1.56].

But the droplet is, in general, not in equilibrium with respect to sulfuric acid. If the partial pressure of sulfuric acid in the environment exceeds the vapor pressure of sulfuric acid in the solution droplet, then more sulfuric acid molecules impinge on the particle than are evaporated from it, and the particle will grow. However, each time a new acid molecule is incorporated into the droplet it disturbs the water equilibrium which can only be reestablished by absorbing (about 2) water molecules. The evaporation is just the inverse of the growth process. In either case, the droplet always maintains equilibrium with respect to water, and it will grow or evaporate depending upon whether the sulfuric acid partial pressure is greater or smaller, respectively, than the sulfuric acid vapor pressure. This process has been described by HAMILL [1.57], HOPPEL [1.58], and HAMILL et al. [1.59].

It should be noted that growth by condensation depends upon the Knudsen number Kn, defined as the ratio of the mean free path of the gaseous molecules to the particle radius. For small particles, $Kn \gg 1$ and the growth rate of the particle (dr/dt) is directly proportional to the rate of impingement of sulfuric acid molecules. For large particles, $Kn \ll 1$ and the growth by diffusion is inversely proportional to the radius of the particle.

The general expression for growth by heteromolecular condensation is [1.59]

$$\frac{dr}{dt} = \frac{\bar{v}D(P_a - P_a^0)/kT}{r\chi(1 + \lambda Kn)} \tag{1.1}$$

where P_a is the partial pressure of H_2SO_4, P_a^0 is the vapor pressure of H_2SO_4 in a solution of concentration χ, D is the diffusion coefficient described by FUCHS and SUTUGIN [1.60], λ is a correction factor depending upon Kn and the sticking coefficient, and $\bar{v}$ is the average volume per molecule in the droplet. The concentration of sulfuric acid in the droplet (χ), the vapor pressure of H_2SO_4 (P_a^0), and the average molecular volume all depend upon the amount of water vapor in the atmosphere and upon the temperature.

An interesting aspect of the growth by condensation is that for a given value of P_a, the growth of particles with radius less than about 0.5 μm is faster at low altitudes whereas the growth of larger particles is faster at higher altitudes.

1.3.3 Coagulation

Because of Brownian motion, a collection of aerosol particles will undergo mutual collisions. A large number of these collisions will result in the coalescence of the two colliding particles, leading to the formation of a single particle whose volume is the sum of the volumes of the two initial particles. The coagulation of particles is one of the most important physical processes affecting the evolution of an aerosol. Unfortunately, it is also the process least amenable to theoretical treatment because it obeys an equation which must be solved by numerical techniques.

The coagulation equation describes the change in the number n_i of particles of volume v_i due to collisions with other particles. If K_{ij} is the probability of a collision between a particle of volume v_i and one with volume v_j, then the number of i particles lost per second due to such collisions is

$$\sum_{j=1}^{\infty} n_i n_j K_{ij} \quad .$$

However, particles of volume v_i are also created by collisions between particles of volumes v_j and v_k (where $v_j + v_k = v_i$). Thus the rate of formation of i particles is

$$\frac{1}{2} \sum_{\substack{j=1 \\ (k=i-j)}}^{\infty} n_j n_k K_{jk} \quad ,$$

where the factor $\frac{1}{2}$ avoids counting each collision twice.

Consequently, the rate of change of n_i is

$$\frac{dn_i}{dt} = - \sum_{j=1}^{\infty} n_i n_j K_{ij} + \frac{1}{2} \sum_{\substack{j=1 \\ (k=i-j)}}^{i} n_j n_k K_{jk} \quad . \tag{1.2}$$

For continuous particle size spectra the summations in (1.2) are replaced by integrals in which $n(v)$ is the number of particles with volume between v and $v + dv$. Then,

$$\frac{dn(v)}{dt} = - \int_0^{\infty} n(v)n(u)K(v,u)du + \frac{1}{2} \int_0^{v} n(u)n(v-u)K(u,v-u)du \quad . \tag{1.3}$$

This integro-differential equation can be solved only for a few special forms of the kernel $K(u,v)$, which are not of particular physical interest. Consequently, the approach generally used for evaluating dn_i/dt is to solve the discrete form of the equation by numerical techniques as discussed in Chap.4.

The literature on the coagulation equation is vast and we will not attempt to present the reader with a long list of references. However, GELBARD et al. [1.61, 62] have recently carried out a number of interesting studies and have developed several promising new approaches to the problem.

1.3.4 Sedimentation

Of the various microphysical processes affecting the aerosol particles, gravitational sedimentation is the simplest to treat. The fall velocity of a particle is determined by evaluating the velocity at which the gravitational force is equal to the sum of the buoyancy force and the resistance force which is a function of Reynolds number [1.63]. For particles smaller than about 80 μm, the terminal velocity for a particle of radius r and density ρ_p is given by Stokes law

$$v_T = C(2/9) \frac{\rho_p - \rho_a}{\eta} gr^2 \tag{1.4}$$

where ρ_a is the air density, η is the viscosity of air, g is the acceleration due to gravity, and C is called the Cunningham factor which is given by

$$C = (1 + K\lambda/r) \ . \tag{1.5}$$

Here, λ is the mean free path of air molecules and K is a constant with value between 0.8 and 0.86. The Cunningham factor is important for particles smaller than about 0.1-μm diameter which describe motions very similar to those of gas molecules.

JUNGE et al. [1.2] and KASTEN [1.64] have thoroughly discussed gravitational sedimentation. It is interesting to note that at 20-km altitude a 0.1-μm particle has a fall velocity of about 5×10^{-3} cm s^{-1}, which is equivalent to stating that over half a year would be required for such a particle to fall one kilometer. At lower altitudes the fall velocity is, of course, even smaller.

1.3.5 Transport

The aerosol layer is strongly influenced by atmospheric motions, which distribute the precursor gases and vapors as well as the particles themselves throughout the atmosphere. For very long-lived gaseous species, the mixing tendencies of the motions lead to near-uniform mixing fractions; a good example is carbonyl sulfide which displays a nearly constant tropospheric mixing ratio all over the globe. On the other hand, sulfur dioxide is not uniformly distributed in either the troposphere or the stratosphere because of its relatively short lifetime. Stratospheric particles must be treated a little differently than the gases because of sedimentation.

In a one-dimensional model simulation [1.36], the vertical transport is simulated by expressing the corresponding particle flux ϕ as a product of the vertical mixing ratio gradient $\partial c/\partial z$ with an "eddy diffusion" coefficient K and the number density of atmospheric molecules n_A,

$$\phi = - K n_A (\partial c/\partial z) - v_T n_A c \tag{1.6}$$

where v_T is the sedimentation velocity discussed in the preceding section. If the constituent in question is a gas, we necessarily have $v_T = 0$. It is important to note that K has very little physical significance other than the tendency toward mixing equilibrium, which states that the flux is roughly proportional to the mixing ratio gradients for the global spatial scale and for long time scales. Extension of an aerosol model to two or three dimensions requires a more elaborate treatment of transport by inclusion of a suitable simulation of horizontal motions. Because such simulations have been carried out only for tracers such as volcanic ash and not for evolving aerosols, we will not consider them further here.

References

1.1 P. Gruner, H. Kleinert: Prob. Kosm. Phys. *10* (1927)
1.2 C.E. Junge, C.W. Chagnon, J.E. Manson: J. Meteorol. *18*, 81-108 (1961)
1.3 P.J. Crutzen: Geophys. Res. Lett. *3*, 73-76 (1976)
1.4 H.H. Lamb: Philos. Trans. R. Soc. London *266*, 425-533 (1970)
1.5 E.K. Bigg: J. Atmos. Sci. *32*, 910-917 (1975)
1.6 G.V. Ferry, H.Y. Lem: In Proceedings of the Third Conference on the Climatic Impact Assessment Program; Rpt. DOT-TSC-OST-74-15, Department of Transportation, Washington, D.C. (1974), pp.310-317
1.7 N.H. Farlow, G.V. Ferry, H.Y. Lem, D.M. Hayes: J. Geophys. Res. *84*, 733-743 (1979)
1.8 A.L. Lazrus, B.W. Gandrud: J. Geophys. Res. *79*, 3424-3431 (1974)
1.9 J.M. Rosen: J. Geophys. Res. *69*, 4673-4676 (1964)
1.10 G. Fiocco, G. Grams: J. Atmos. Sci. *21*, 323-324 (1964)
1.11 L. Elterman, R.B. Toolin, J.D. Essex: Appl. Opt. *12*, 330-337 (1973)
1.12 P.B. Russell, W. Viezee, R.D. Hake, R.T.H. Collis: Q. J. R. Meteorol. Soc. *102*, 675-695 (1976)
1.13 M.P. McCormick, T.J. Swissler, W.P. Chu, W.H. Fuller: J. Atmos. Sci. *35*, 1296-1303 (1978)
1.14 M.P. McCormick, P. Hamill, T.J. Pepin, W.P. Chu, T.J. Swissler: Bull. Am. Meteorol. Soc. *60*, 1038-1046 (1979)
1.15 C.E. Junge, J.E. Manson: J. Geophys. Res. *66*, 2163-2182 (1961)
1.16 E.K. Bigg, A. Ono, W.J. Thompson: Tellus *22*, 550-563 (1970)
1.17 A.L. Lazrus, B. Gandrud, R.D. Cadle: J. Geophys. Res. *76*, 8083-8088 (1971)
1.18 N.H. Farlow, D.M. Hayes, H.Y. Lem: J. Geophys. Res. *82*, 4921-4929 (1977)
1.19 D. Hayes, K. Snetsinger, G. Ferry, V. Overbeck, N. Farlow: Geophys. Res. Lett. *7*, 974-976 (1980)
1.20 T.J. Pepin, M.P. McCormick, W.P. Chu, F. Simon, T.J. Swissler, R.R. Adams, K.H. Crumbly, W.H. Fuller: "Stratospheric Aerosol Measurements;" Rpt. NASA SP-421, NTIS, Springfield, Virginia (1977) pp.127-136
1.21 D.J. Hofmann, J.M. Rosen, T.J. Pepin, R.G. Pinnick: J. Atmos. Sci. *32*, 1446-1456 (1975)
1.22 J.M. Rosen, D.J. Hofmann, J. Laby: J. Atmos. Sci. *32*, 1457-1462 (1975)
1.23 V.R. Overbeck, N.H. Farlow, G.V. Ferry, H.Y. Lem, D.M. Hayes: Geophys. Res. Lett. *8*, 15-17 (1981)
1.24 P.B. Russell, R.D. Hake: J. Atmos. Sci. *34*, 163-177 (1977)
1.25 N.D. Sze, M.K.W. Ko: Nature London *278*, 731-732 (1979)
1.26 N.D. Sze, M.K.W. Ko: Nature London *280*, 308-310 (1979)
1.27 J.A. Logan, M.B. McElroy, S.C. Wofsy, M.J. Prather: Nature London *281*, 185-188 (1979)
1.28 A.R. Ravishankara, N.M. Kreutter, R.C. Shah, P.H. Wine: Geophys. Res. Lett. *7*, 861-864 (1980)
1.29 R.S. Iyer, F.S. Rowland: Geophys. Res. Lett. *7*, 797-800 (1980)
1.30 R.P. Turco, R.C. Whitten, O.B. Toon, J.B. Pollack, P. Hamill: Nature London *283*, 283-286 (1980)
1.31 W. Jaeschke, R. Schmitt, H.-W. Georgii: Geophys. Res. Lett. *3*, 517-519 (1976)
1.32 H.-W. Georgii, F.X. Meixner: J. Geophys. Res. *85*, 7433-7438 (1980)
1.33 E.C.Y. Inn, J.F. Vedder, D.O'Hara: Geophys. Res. Lett. *8*, 5-8 (1981)
1.34 E.C.Y. Inn, J.F. Vedder, B.J. Tyson, D. O'Hara: Geophys. Res. Lett. *6*, 191-193 (1979)
1.35 A.L. Torres, P.J. Maroulis, A.B. Goldberg, A.R. Bandy: J. Geophys. Res. *85*, 7357-7360 (1980)
1.36 R.P. Turco, P. Hamill, O.B. Toon, R.C. Whitten, C.S. Kiang: J. Atmos. Sci. *36*, 699-717 (1979)
1.37 J.W. Burgmeier, I.H. Blifford: Water, Air, Soil Pullution 5, 133-147 (1975)
1.38 M.A. Kritz: "Formation Mechanism of the Stratospheric Aerosol"; Ph. D. Dissertation, Yale University (1975)
1.39 M.A. Kritz: J. Phys. Paris *36*, Coll. C8, Suppl. 12, 17-23 (1975)
1.40 J.M. Rosen, D.J. Hofmann, S.P. Singh: J. Atmos. Sci. *35*, 1304-1313 (1978)
1.41 R.C. Whitten, O.B. Toon, R.P. Turco: Pure Appl. Geophys. *118*, 86-127 (1980)

1.42 J.B. Pollack, O.B. Toon, A. Summers, B. Baldwin, C. Sagan, W. Van Camp: Nature London *263*, 551-555 (1976)
1.43 J.B. Pollack, O.B. Toon, C. Sagan, A. Summers, B. Baldwin, W. Van Camp: J. Geophys. Res. *81*, 1071-1083 (1976)
1.44 J.B. Pollack, O.B. Toon, A. Summers, W. Van Camp, B. Baldwin: J. Appl. Meteorol. *15*, 247-258 (1976)
1.45 R.P. Turco, O.B. Toon, J.B. Pollack, R.C. Whitten, I.G. Poppoff, P. Hamill: J. Appl. Meteorol. *19*, 78-89 (1980)
1.46 W.H. Marlow: *Aerosol Microphysics I*, Topics in Current Physics, Vol.16 (Springer, Berlin, Heidelberg, New York 1980)
1.47 E.A. Martell: Tellus *28*, 486-498 (1966)
1.48 G.M. Hidy, J.L. Katz, R. Mirabel: Atmos. Environ. *12*, 887-892 (1978)
1.49 P. Hamill, C.S. Kiang, R.D. Cadle: J. Atmos. Sci. *34*, 150-162 (1977)
1.50 P. Hamill, R.P. Turco, O.B. Toon, C.S. Kiang, R.C. Whitten: J. Aerosol Sci. (in press, 1982)
1.51 F. Arnold, R. Fabian, W. Joos: Geophys. Res. Lett. *8*, 293-296 (1981)
1.52 V.A. Mohnen: In Proceedings of the Fourth Conference on the Climatic Impact Assessment Program; Rpt. DOT-TSC-OST-75-38, Department of Transportation, Washington, D.C. (1975), pp.478-491
1.53 R.P. Turco, O.B. Toon, P. Hamill, R.C. Whitten: J. Geophys. Res. *86*, 1113-1128 (1981)
1.54 J.P. Friend, R.A. Barnes, R.M. Vasta: J. Phys. Chem. *84*, 2423-2436 (1980)
1.55 F. Arnold: Nature London *284*, 610-611 (1980)
1.56 J.M. Rosen: J. Appl. Meteorol. *10*, 1044-1045 (1971)
1.57 P. Hamill: J. Aerosol. Sci. *6*, 475-482 (1975)
1.58 W.A. Hoppel: J. Rech. Atmos. *9*, 67-180 (1976)
1.59 P. Hamill, O.B. Toon, C.S. Kiang: J. Atmos. Sci. *34*, 1104-1119 (1977)
1.60 N.A. Fuchs, A.G. Sutugin: "High Dispersed Aerosols," in *Topics in Current Aerosol Research*, Vol.2, ed. by G.M. Hidy, J.R. Brock (Pergamon, New York 1971)
1.61 F. Gelbard, J.H. Seinfeld: J. Comp. Phys. *28*, 357-375 (1978)
1.62 F. Gelbard, Y. Tambour, J.H. Seinfeld: "Sectional Representations for Simulating Aerosol Dynamics" (Unpublished, 1980)
1.63 H.R. Byers: *Elements of Cloud Physics* (University of Chicago Press, Chicago 1965)
1.64 F. Kasten: J. Appl. Meteorol. *7*, 944-947 (1968)

Additional Reference
=====

A.A. Viggiano, F. Arnold: "Extended sulfuric acid vapor concentration measurements in the stratosphere". Geophys. Res. Lett. *8*, 583-586 (1981)

2. Observations

E.C.Y. Inn, N.H. Farlow, P.B. Russell, M.P. McCormick, and W.P. Chu

With 27 Figures

The data which are critical to any physical phenomenon are the results of observations. Hence, this chapter is fundamental to the entire book. Although in Chap.1 we briefly touched on the area of field measurements, we will treat the techniques and results much more exhaustively here. It is appropriate to classify the sections of the chapter according to the objects of measurement and the techniques employed. First we discuss the in situ measurement of precursor gases such as carbonyl sulfide, then in situ sampling and analysis of aerosol particles, followed by ground-based remote sensing of the aerosol layer by lidar, and finally satellite remote sensing of the layer.

2.1 Measurements of Precursor Gases in the Stratosphere

Photochemical processes of the sulfur cycle in the stratosphere leading to the formation of sulfate aerosols involve the stable, reduced sulfur gases carbonyl sulfide (OCS), carbon disulfide (CS_2), and sulfur dioxide (SO_2). The presence of these precursor gases, their concentrations, and their distribution in the stratosphere set important constraints on the formation and distribution of sulfate aerosols. Only recently have measurements been made in the stratosphere of mixing ratios of these precursor gases. Although the number of measurements reported thus far is limited, they provide important background data on the presence and distribution of these constituents. The experimental techniques used in stratospheric measurements of these precursor gases will be described and recent results will be summarized.

2.1.1 In Situ Mass Spectrometer Method

Clearly, in situ stratospheric measurements require the use of airborne platforms to support on-board equipment for analysis at high altitude. High-flying jet aircraft have been used as platforms in sampling experiments of precursor gases in the lower stratosphere (from tropopause to about 21 km) and balloons for measurements up to about 30 km.

---

* Sections 2.1-3 of this chapter were written by the authors in the order they are listed; the last two are jointly responsible for Sect.2.4

The earliest measurements of stratospheric sulfur gases from a balloon platform were reported as recently as 1977 by SAGAWA and ITOH [2.1]. The analytical system included a small magnetic sector mass spectrometer with an electron impact source and a channel electron multiplier detector. A minimum detectable concentration of about 0.1 ppmv (parts per million by volume) is quoted for any species not in the background of the instrument. Measurements were carried out on 25 May, 1975 during descent of the balloon from 30 to 10 km. Large signals were observed at mass peak 64 amu which were assumed to be due to SO_2; the concentrations derived from these varied from 30 to 10 ppmv at altitudes from 16 to 30 km. These are extremely large mixing ratios for SO_2 in the stratosphere, greater than 10^5 times the recently reported value of 0.05 ppbv by JAESCHKE et al. [2.2]. SAGAWA and ITOH [2.1] have suggested that volcanic emission from the 1974 eruption of Mt. Fuego may have contributed to the observed enhancement of the SO_2 concentration. However, the large signals observed may be spurious in that SO_2 may have been produced from local sulfur sources in the spectrometer system, e.g., accumulation of sulfate aerosols deposited in the inlet system or sulfur sources carried aloft in the balloon system. In the concluding remarks by these investigators they stated that the large mixing ratios of SO_2 deduced from the signals should be related to total sulfur rather than to stratospheric SO_2.

2.1.2 In Situ Filter Collection

A major problem encountered in adapting collection methods for measuring minor constituents in stratospheric gases is related to the chemical stability of the gases in the collector. One way of avoiding this problem is to immobilize the constituents during the collection process, thus maintaining the chemical integrity of the collected constituent. Immobilization may be effected by chemically combining each stratospheric constituent with a specific reactant. During analysis the immobilized constituent is chemically released and detected quantitatively.

A technique of immobilizing SO_2 using a filter method was first described and used by JAESCHKE et al. [2.2] and later by GEORGII and MEIXNER [2.3] for stratospheric measurements of this sulfur constituent. The technique is based on the reaction between SO_2 and TCM (tetrachloromercurate: Na_2HgCl_4), in which SO_2 is bound to TCM to form the dichlorosulfitomercurate complex. Analysis for SO_2 is carried out by oxidizing the complexed SO_2 to SO_4^{2-} in an acid solution of $KMnO_4$. STAUFF and JAESCHKE [2.4] have shown that this reaction is accompanied by chemiluminescence wherein the light yield is found to be proportional to the complexed SO_2. The collectors are prepared by impregnating the filter material (Delbag microsorban 98) with a solution of 0.1-$\underline{M}$ TCM. Stratospheric air is then drawn through the filter at flow rates of about 10 STP l/min until a predetermined amount of air is sampled. The complexed SO_2 is recovered from the filter by washing it with the TCM solution. Analysis for SO_2 is then conducted by the chemiluminescence method.

In flights at and above tropopause altitudes the air was sampled from the cabin pressurizing system rather than directly from outside the aircraft. This was found to be acceptable inasmuch as good agreement was obtained in SO_2 measurements in flight experiments involving a second aircraft sampling the outside air directly during flights by both aircraft in the same air mass. The volume of air sampled in all flight measurements was limited to about 100 to 200 STP l. GEORGII and MEIXNER [2.3] have shown that the collection efficiency of the filter, determined by measurements with 0.105- and 0.4-ppbv SO_2 samples prepared by dynamic dilution, was 100%, and remained constant to within 10% for volumes up to 200 STP l. Sampling under these conditions, the limit of detection of stratospheric SO_2 is 0.01 ± 0.001 ppbv.

2.1.3 In Situ Cryogenic Collection

Another sampling method has been developed and used extensively for measurement of minor constituents in the stratosphere, such as $CFCl_3$, CF_2Cl_2, CCl_4, N_2O, CO_2 by VEDDER et al. [2.5] and OCS by INN et al. [2.6]. The method consists of flowing stratospheric air through a liquid-nitrogen-cooled cryogenic collector in which all condensable constituents are simultaneously trapped. Sampling from the same air mass and simultaneous collection of the minor constituents provide a basis for comparison of the concentrations of the trapped gases. The condensed constituents are immobilized at liquid-nitrogen temperatures and may be stored in this condition until ready for analysis.

A major problem encountered in the use of metal containers for trapping and storing stratospheric minor constituents is the stability of these gases in the container. In particular, low concentrations of the precursor sulfur gases (~ few parts per million and lower) cannot be stored at room temperature in metal containers without substantial decay in concentration occurring in a period of several hours. By carefully passivating the interior surfaces of the metal container, chemisorption of the gases on the walls may be avoided, as has been done by SCHMELTEKOPF et al. [2.7] for sampling stratospheric halocarbons. INN et al. [2.6], however, found the use of glass collectors for trapping and storing sulfur gases (and other stratospheric minor constituents) to be quite satisfactory. Analysis of gas samples stored in a glass container at room temperature over a period of a few days showed no discernible decay in concentration.

The cryogenic collector, as shown in Fig.2.1, consists of pyrex tubing, about 1.7-cm inside diameter and 1 m long, shaped in the form of a helix with many deep dimples throughout the tubing. Each end of the tubing is terminated by stainless steel valves, and the volume of the collector is about 200-300 ml. Sampling is carried out by flowing air through the glass collector, which is mounted in a dewar and immersed in liquid nitrogen. Generally, about 200-300 STP l of stratospheric air are sampled and the concentration of the sulfur constituents is enriched over

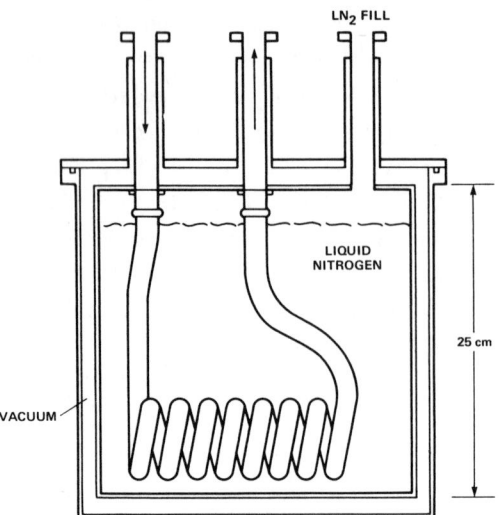

Fig.2.1
Glass cryogenic collector [2.8]

1000 times in the collector. Analysis of samples for sulfur gases is carried out by gas chromatography using a flame photometric detector.

In early stratospheric experiments reported by INN et al. [2.6] and VEDDER et al. [2.5,8] the trapping efficiency of glass cryogenic collectors was determined during sampling in flight by connecting two collectors in series. Trapping efficiency is defined as the ratio of moles of constituent retained in the trap to the total entering the trap. In the experiments of VEDDER et al. [2.8] the average trapping efficiency for constituents $CFCl_3$, CF_2Cl_2, N_2O, and CO_2 was 0.86 ± 0.04 at flow rates ranging from about 12 to 15 STP l/min. A much larger scatter in trapping efficiency was observed for the OCS measurements reported by INN et al. [2.6], ranging from about 0.44 to 0.92 for flow rates 2.4 to 12 STP l/min. Much of this scatter can be attributed to sample-handling technique and gas chromatographic analysis of OCS. In later flight experiments, with improved technique and better control of the sampling conditions, the trapping efficiency was 0.87 ± 0.12 as reported by INN et al. [2.9]. These experiments, as well as laboratory studies of trapping efficiency of the glass cryogenic collectors, tend to indicate that, within the scatter associated with the precision of the analytical measurements, trapping efficiency is the same for all stratospheric minor constituents whose boiling points are greater than about -100 °C for the same trapping conditions of pressure, flow rate, and total volume sampled (e.g., less than about 300 STP l).

The sample-handling technique of the cryogenically collected gases is as follows. After sampling, residual air in the collector is removed by evacuation with a sorption pump while the condensed gases are still at liquid-nitrogen temperature. The sample now consists almost totally of condensed CO_2 and small amounts of the minor stratospheric constituents. The condensate is then distilled and transferred into a glass bulb at liquid-nitrogen temperature. The glass bulb (about 200-250 ml) is fitted with a glass valve so that the sample is now stored in an all-glass container. Furthermore, the sulfur constituents may now be immobilized by storing the sample bottle in liquid nitrogen, thus maintaining the chemical integrity of the sample.

Analysis of the stratospheric samples is performed by conventional gas chromatography. Separation of the sulfur constituents is carried out under isothermal conditions (about 35-55ºC) in teflon columns packed with either Porapak Q or Chromosil-310. Temperature programming is also used to alter the separation of the gases. The sulfur constituents are detected with a flame photometric detector (390-nm sulfur filter). Carbonyl sulfide may also be detected and measured with a photoionization detector and an 11.2-eV light source. In order to minimize any degradation of the sulfur gases due to prolonged contact with metal, teflon tubing is used as the sample injection loop. Except for the stainless steel multiport valve for injection of the sample into the gas chromatograph, the rest of the transfer manifold is made of glass. Thus, rapid transfer is effected to reduce the residence time of the sample in the metal sections.

In analysis of the samples, signals of the stratospheric gases are always bracketed by those of calibration standards. The latter were prepared in 1-l glass flasks at pressures of about 1 atmosphere by mixing pure gases consisting of the sulfur constituent and CO_2. The concentration of the calibration mixture is determined from the measured partial pressure of each constituent, based on 0.1% accuracy pressure gauges. Calibration standards are prepared with concentrations from about one to a few ppmv, which are approximately in the concentration range expected to be encountered in the enriched trapped samples. The sulfur constituents remain very stable in the glass flasks with no discernible decay in concentration over a period of several weeks. Intercomparison of these primary standards with commercially prepared standards provide a calibration set consistent to within about 5%. These two types of standards have been used interchangeably during analysis of the stratospheric sample.

Flight experiments have been carried out mainly in the lower stratosphere in the sulfate aerosol layer. The NASA U-2 aircraft has been used as the platform for the stratospheric cryogenic sampling system for flights above the tropopause up to altitudes of about 21 km. One balloon-borne measurement has been made of OCS at an altitude of 31.2 km, using the same cryogenic sampling system [2.6]. The gondola carrying the sampling system is deployed about 300 m below the balloon to avoid contamination due to outgassing from the balloon. A 2000-l capacity sorption pump is used to draw air through the cryogenic collectors. The intake port of the sampling system is oriented into the wind by a sail mounted on the gondola.

Figure 2.2 is a diagram of the flight configuration of the sampling system. The latter (payload of about 225 kg) is mounted in the instrument bay of the aircraft. As shown in Fig.2.2, the system consists basically of an inlet and an outlet mani-

CRYOGENIC SAMPLING SYSTEM

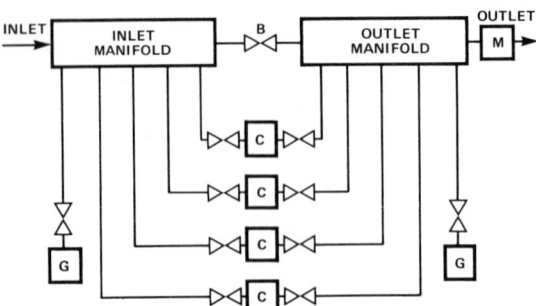

Fig.2.2. Diagram of cryogenic sampling system; (B) all metal high-vacuum bypass valve; (C) glass collector; (G) whole air sampler; (M) venturi mass flow meter

fold, four cryogenic collectors (C), and a venturi mass flow meter (M). Also shown in the diagram are whole air samplers (G) for collecting stratospheric air at ambient pressures. The inlet and outlet parts (not shown in Fig.2.2) are located outside the boundary layer of the skin of the aircraft. Flow of air through the system is driven by ram pressure. All parts of the system are made of stainless steel. System preparation for a flight experiment consists of heating each cryogenic sampler separately to 120°C while purging it with clean nitrogen for several hours until the background level of any out-gassing is just below the limit of detection.

Each sample is acquired during flight at a predetermined height with the aircraft flying at a constant speed of about 700 km/hr and on a constant pressure altitude flight track. Prior to collection of each sample, the bypass valve, B (Fig.2.2), is opened to purge the system with ambient air at flow rates of about 30 STP l/min for several minutes. The flow of air is then directed to a collector by closing B and opening the two valves of the collector. Each valve is driven by a small electric motor. The sampling flow rates are adjusted to about 10 to 15 STP l/min and the amount of air sampled to about 200 to 300 STP l. At these flow rates, the residence time of the sampled air in the system is ≤ 1 s so that no measurable decay is expected. This is consistent with laboratory studies of the cryogenic trapping method. For samples collected as described above, the lower limit of measurement of the stratospheric mixing ratio is a few pptv.

Because the vapor pressure of ozone at liquid-nitrogen temperature is about 0.1 Torr, which is very much greater than its partial pressure in the lower stratosphere, it is not expected that any ozone will be trapped during the sampling process. On the other hand, trapping of water vapor may introduce some problems when large volumes of air are sampled. Owing to enrichment occurring in the trapping process, liquid water may be formed in the collector if the partial pressure of the collected water exceeds its vapor pressure at room temperature. Limiting the total volume of air sampled clearly prevents this from occurring.

2.1.4 Remote Sensing Methods

The only stratospheric sulfur constituent that has been measured by a remote-sensing technique is OCS. MANKIN et al. [2.10] reported infrared spectroscopic measurements of OCS in the stratosphere using a Fourier transform spectrometer mounted on board the NCAR Sabreliner aircraft. The OCS overburden is measured from the absorption spectra obtained when flying at about 12-km altitude.

The spectroscopic method is based on OCS absorption in the ν_3 band at 2062 cm^{-1}. There is partial overlap with the $\nu_1 + \nu_3$ band of O_3 at 2110 cm^{-1} and the $\nu_1 + \nu_2$ band of CO_2 at 2077. Measurement from above the upper troposphere minimizes unwanted absorption by water vapor in the lower atmosphere. Extensive laboratory measurements of the spectra of OCS with the same spectrometer used in the flight measurements were carried out under a variety of conditions. From these measurements a band strength of ν_3 is derived, which is used for the analysis of the flight spectra.

Spectra were obtained in flights near sunrise and sunset, taking advantage of the long optical path of sunlight at these times, thus enhancing the absorption of the weak OCS lines. Spectra were also taken over a range in latitude from about 0º to 40ºN. The OCS abundance was determined from an analysis of the spectrum in the region 2050-2060 cm^{-1}. Column abundances were determined from comparison of the observed spectra with synthetic spectra generated from assumed OCS profile distributions, measured line strengths, constant collision line width of 0.12 cm^{-1}, and varying OCS abundance.

2.1.5 Summary of Stratospheric Observations of Sulfur Gases

Only a small number of measurements of precursor sulfur gases in the unperturbed stratosphere have been reported. During the recent eruption of Mt. St. Helens on 18 May 1980 a number of measurements of sulfur gases made in the plume which penetrated the tropopause as reported by INN et al. [2.11]. These measurements of the perturbed stratosphere will not be considered in this summary.

All results of sulfur gas measurements in the stratosphere will be compared with those presented in Fig.2.3, which is taken from the report by INN et al. [2.9]. Although there appears to be appreciable scatter in the OCS data, the shape of the profile is evident. Some of the scatter is undoubtedly due to experimental uncertainties. However, it is possible that natural variability of the concentrations of minor constituents in the stratosphere may also contribute significantly to the observed scatter. Such variability has been observed by GOLDAN et al. [2.12]. Apparent latitude dependence of the profile of $CFCl_3$, CF_2Cl_2, and N_2O has been observed in stratospheric measurements by VEDDER et al. [2.5]. Similar latitude dependence may also account for part of the observed scatter in the OCS data. Also shown in Fig.2.3 is the predicted profile of OCS deduced from one-dimensional model calculations of TURCO et al. [2.13]. The results of MANKIN et al. [2.10] yield a

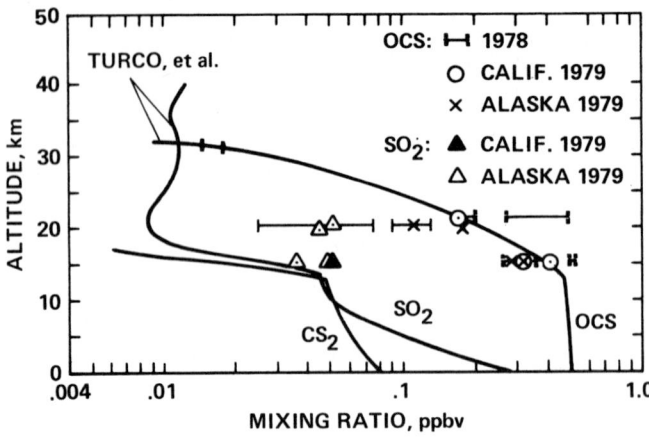

Fig.2.3. Altitude distribution of sulfur gases [2.9]. Curves are profiles predicted by [2.13]

a similar profile of OCS, deduced from their spectroscopic measurements. The tropospheric mixing ratio of about 0.50 ppbv and the extension of the OCS profile into the troposphere are based on the measurements of MAROULIS et al. [2.14] and SANDALLS and PENKETT [2.15].

Measurements of SO_2 in the lower stratosphere were first reported by JAESCHKE et al. [2.2] in which a mixing ratio of 0.050 ppbv at 13 km was obtained. Continuing this effort, GEORGII and MEIXNER [2.3] reported SO_2 measurements in the stratosphere obtained during six flight experiments in various locations in Europe. Mixing ratios of about 0.01 to 0.1 ppbv were measured at altitudes from about 10 to 15 km. The SO_2 results plotted in Fig.2.3 are in general agreement with the above, indicating that in the lower stratosphere the mixing ratio of SO_2 lies somewhere in the range of 0.01 to 0.1 ppbv. Because of the scatter and variability of the data, it is not possible to discern any trend in the profile in this region of the stratosphere. The SO_2 results of SAGAWA and ITOH [2.1] are clearly not compatible with the lower stratosphere measurements. The SO_2 profile predicted by model calculations by TURCO et al. [2.13], as shown in Fig.2.3, and that predicted by SZE and KO [2.16] both appear to fall short of the observed mixing ratios, especially near the region of the aerosol maximum at about 18-20 km. TOON et al. [2.17] have recently modified and updated the model calculations of TURCO et al. [2.13] with predicted profiles in better agreement with the SO_2 data.

At the present time, the existence of CS_2 in the stratosphere remains uncertain. In the flight experiments conducted by INN et al. [2.9] they reported that during the gas chromatography analysis of all the 15.2-km samples, weak signals were observed at retention times at which CS_2, if present, should appear. Furthermore, these weak signals were not apparent in the chromatograms of the higher altitude samples. This latter observation would tend to indicate that the observed weak signals were not due to contamination or instrumental artifact. These observations suggest that, if the weak signals are due to CS_2, an upper limit mixing ratio is estimated to be about 0.001 ppbv with an uncertainty of a factor of 3-4. This low concentration of CS_2 appears to be consistent with the low mixing ratio of 0.030 ppbv of CS_2 in the troposphere as reported recently by MAROULIS and BANDY [2.18]. TOON et al. [2.17] have considered these recent CS_2 data and have thus suggested modifications of the TURCO et al. [2.13] profile shown in Fig.2.3.

2.2 Aircraft and Balloon Measurements of Aerosol Properties

As was pointed out in Chap.1, the nature of the particles normally present in the stratospheric aerosol as well as those due to the occasional injection of volcanic products is of interest because these aerosols may affect climate. Aircraft and balloon sampling over the past two decades [2.19] have established the composition of the aerosols, which is the main subject of this section.

2.2.1 Aerosol Constituents

Stratospheric aerosols are composed of variable mixtures of Aitken nuclei, larger acid droplets (sometimes containing granular inclusions), cosmic dust, rocket exhaust products, and volcanic ash. The tiniest particle constituents (Aitken nuclei, ≤ 0.1 μm diameter) have been studied by special methods that cause them to grow to larger, more observable sizes. Larger acid droplets (0.1 to 1 μm), usually comprising most of the aerosol, have also been examined with a variety of detectors and collecting instruments during volcanic periods and quiet times. Scientists have, however, infrequently analyzed larger acid droplets for granular inclusions, believed to be of cosmic origin, or of tropospheric or volcanic derivation; and only a few investigators have collected and examined large (≥ 3 μm), solid grains of cosmic dust that enter the atmosphere as primordial particles or ablate from larger meteors. Along with these particles, aluminum oxide spherules of similar size derived from exhausts of solid-fuel rockets are often found. After major volcanic explosions, intrusions of ash, with grain sizes from about 0.1-μm diameter to as large as 30 μm, have been especially carefully studied.

2.2.2 Aerosol Properties

Those particle properties most likely to determine how aerosols scatter and absorb energy from the sun and Earth are particle size, shape, composition, and physical state. Where the constituents of particles originate, how the particles form and grow to maturity, and how and where they leave the stratosphere determine the time span of their effects. Together, all of these features affect upper atmospheric heating or cooling and ground-level temperature changes.

2.2.3 Detection Methods

Scientists have measured aerosol properties by placing special sensors and collectors on research aircraft and balloons. The sensor instruments generally record aerosol properties without physically collecting particles for later laboratory analysis. In contrast, collectors return samples to Earth for study.

Sensors used in stratospheric applications mainly detect intensity of light scattered from individual particles passing through a light beam inside the unit. Numbers of particles associated with various scattering intensities from a known volume of air produce desired particle size distributions. Some Aitken nuclei counters are special adaptations of this counting and sizing principal. Nuclei are too small to be measured directly because of the 0.1 μm diameter threshold sensitivity of this detector method. Therefore, detector systems first condition nuclei in a supersaturated chamber or expansion cylinder where vapor condenses on nuclei and they grow to measurable dimensions. Fluids such as water or ethylene glycol fulfill this purpose [2.20]. Other nuclei counting systems photograph droplets or detect acoustic signals of droplets passing through capillaries to determine nuclei concentrations [2.20]. All of these nuclei counters determine concentrations, but not initial nuclei sizes.

Collectors are used mostly in the stratosphere to obtain those acid droplets and solid grains larger than about 0.1 μm diameter. These particles are analyzed in the laboratory for sizes, concentrations, physical properties, and chemical composition. Two methods of collection are usually employed: inertial impaction and filtration.

Inertial impaction includes both jet impaction and airstream collection. Jet impactors suck in the airstream through small orifices or slits to impinge it against collecting plates. Particles of a certain critical size and larger cross the diverging airstream to deposit on the plate, while smaller particles follow the stream around the collector. The airstream cascades through a series of smaller and smaller orifices and particles of diminishing mean size collect at later stages, providing rough size cuts. Both single- [2.19,21] and multiple-stage impactors [2.22] are used, some with rotating disk sampling plates [2.23] to time-sequence collections, and others with vibrating quartz crystal plates for mass determinations deduced by changes in crystal vibration frequency [2.24].

Airstream impactors place collecting surfaces directly into the airstream beyond influence of aircraft structures, or into pumped air flows ducted into balloon gondolas. In the aircraft case, high impaction velocities result from aircraft movement through the air [2.25,26,27]; in the balloon case [2.28,29], from on-board pumps. These instruments achieve high collection efficiencies because collecting surfaces are small and air movement is fast [2.30]. In one aircraft configuration, palladium wires (75 μm diameter) are used for collectors that are later viewed in scanning electron microscopes [2.31]. In other cases, fingerlike probes of small cross-sectional area are employed [2.32]. In certain instances electron microscope screens (3 mm diameter) are attached to thin wires and other surfaces in aircraft and balloon-borne systems for later microscopic analysis [2.29,33].

In filter collectors, several varieties of filter media (some impregnated with reagents to detect gases) are used for aerosol sampling [2.34-36]. This collection system provides large samples for classical wet chemical analyses. Voluminous data have been obtained over past decades on worldwide concentrations of important anions and cations in the stratosphere with this method [2.37-39].

2.2.4 Analytical Techniques

Automated wet chemical procedures, combined with neutron activation analyses, provide accurate quantitative assessments of sulfates, nitrates, halides, ammonium ions, and other trace constituents in bulk filter samples [2.37,39]. Required, however, are large sample quantities not easily provided by impaction devices. More qualitative approaches, therefore, are applied to small amounts obtained in jet and airstream impactors. Indeed, in some cases investigators have devised unique approaches to analyze individual particles isolated on collecting surfaces. Among these are in situ reagent-film processes whereby chemicals in collecting film surfaces react directly with particles as they impact [2.40,41]. One example is the reaction of sulfuric acid droplets in the stratosphere with absorbing films containing barium ions that yield a granular reaction spot of barium sulfate. The quantity of acid in the droplet is estimated by assessing reaction spot size. But more information is gained in the laboratory using a battery of microbeam analytical instruments.

Electron microscopes — both transmission and scanning types — use electron microbeams to obtain images that yield particle size and morphology. Moreover, when these instruments are equipped with auxiliary detectors, additional information is obtained about the crystalline composition of particles using electron diffraction [2.42] and elemental makeup by analyzing X-ray emissions [2.43]. Electron spectroscopy for chemical analysis (ESCA), Auger electron methods, and scattered ion mass spectrometry (SIMS) can provide surface composition of some larger grains. In the last method, instruments operate with microbeams composed of ions rather than electrons. This is also the case with ion and proton microprobes whereby specialized analyses can increase chemical knowledge about elements in individual aerosol particles. The Raman microprobe is a newly developed instrument used to qualitatively identify chemical compounds within the particle matrix [2.44]. For instance, in tropospheric aerosols graphitic carbon, sulfuric acid, ammonium sulfate, and other compounds, also significant in the stratosphere, can be detected [2.45].

2.2.5 Statistical Treatments

Particle size distributions are an important product of both light-scattering sensors and collectors. Some sensors provide number counts for all sizes of particles in a range from 0.1 μm to about 3 μm diameters [2.46,47]. Others provide only size groupings such as all particles larger than 0.3 μm diameter and all greater than 0.5 μm [2.48]. In this case, scientists usually report ratios of the two size groups, or concentrations greater than a stated size versus altitude. Collectors commonly provide complete number counts for all sizes of particles observable on the collecting surfaces.

For stratospheric aerosol size distributions like these, two graphical methods of data presentation are usually used. In these, abscissas generally represent particle size (diameter or radius). Ordinates describe particle concentration parameters such as dN/dr cm^{-3} air (number of particles per radius interval per unit volume of sampled air) or $dN/d\log(r) cm^{-3}$ air [2.49]. Number distributions do not show much detail in the size distributions for larger particles because of the rapid drop off of numbers with increasing size. To better see the relationship of various size ranges to integral properties like total particle number, volume, or surface area, plots are best presented as the integral property (divided by $d\log r$) on the ordinate with linear scale and the log of particle size on the abscissa. Thus modes in the distributions and relative numbers of particles, surface areas and volumes in different size ranges are more easily observed.

Investigators believe that multiple modes in size distribution curves occur when young, growing aerosols intermix with older, more mature ones or with particles of different origins [2.50]. Particle data plotted as surface area or volume versus size, as just described, highlight these modes [2.51]. Surface area plotted versus particle size emphasizes modes in the small-particle region because larger surface areas are associated with an equal number of small particles compared to larger ones. Conversely, volume plotted against size enhances large particle modes because relatively bigger volumes are associated with larger grains.

Many other ways of presenting particle concentrations versus various parameters are in use such as concentration cm^{-3} air (ambient or STP) versus altitude, latitude, or other variables. Mixing ratio notations (for example, concentration mg^{-1}) are also useful for comparing one data set with another.

Errors in data are treated in different ways, depending upon uncertainties introduced by particular sampling techniques. For instance, critical features affecting accuracy of light-scattering sensor systems are stability of light sources, sensitivity of photoelectron detectors, and chemical makeup of particles, which affects the index of refraction and, thus, the magnitude of scattered light [2.52]. On the other hand in impactor systems, errors may result from collection inefficiencies, measurement uncertainties in electron microscopes, and physical properties of the particles that affect derivation of sizes [2.21,51]. And in filter systems, important inaccuracies may arise from analytical chemistry processes [2.37]. Of course, other factors common to all these systems influence the experimental accuracy, such as: how representative of the real population is the sample obtained? How effective and accurate are calibration and analytical procedures? Are frequency, time, and locations of the sampling adequate to describe the investigated phenomenon? Quantitative answers are usually elusive. Thus error bars placed on graphical results often do not describe all uncertainties. Nevertheless, analysts have made numerous honest attempts over the past decades to estimate at least qualitative bounds on errors inherent in these difficult measurements.

2.2.6 Aitken Nuclei

CADLE and KIANG [2.20] recently reviewed methods and results of measuring Aitken nuclei in the atmosphere. They concluded that nuclei found below 20 km in altitude are probably of tropospheric origin and are composed of impure sulfuric acid. They suggest that meteors may contribute significant numbers of nuclei above 20 km, an idea consistent with theoretical studies by HUNTEN et al. [2.53]. ROSEN et al. [2.54], using two different Aitken nuclei counters at some locations, surveyed concentrations of these particles from pole to pole in 1976 and 1977. Table 2.1 shows typical stratospheric nuclei mixing ratios and equivalent concentrations at the poles and midlatitudes. These tabular averages were taken from graphical representations [2.54] that show considerable scatter in mixing ratios at all altitudes and locations. Concentrations were calculated from their mixing ratios with use of the *U.S. Standard Atmosphere* Tables [2.55] to calculate cm^3 from their stated air mass and altitude. ROSEN et al. [2.54] characterized polar stratospheric air as having low nuclei mixing ratios (see Table 2.1), while midlatitude air displays relatively high mixing ratios. But in the stratosphere above 20 km, all locations have about the same concentrations of nuclei, consistent with a possible meteoritic source. These authors obtained the nuclei measurements described in Table 2.1 at a time when volcanic disturbances were minimal, the most recent eruption having been Volcán de Fuego in November 1974 about two years earlier. Although no firm evidence exists that eruptions do not introduce large quantities of Aitken nuclei, soundings made within a few months after such eruptions show little change in nuclei content in the stratosphere [2.20,54,56,57]. Therefore, values in Table 2.1 are probably fairly typical of usual concentrations in the upper atmosphere.

2.2.7 Larger Acid Droplets

TOON and FARLOW [2.58], TOON and POLLACK [2.59], and CADLE and GRAMS [2.60] have reviewed the properties of aerosols in the stratosphere. They find most larger acid particles are sulfuric acid droplets [2.61,62] formed by gas-to-particle conversion of sulfur gases [2.63]. Despite many measurements, however, using different techniques over the past two decades, researchers cannot agree on typical size

Table 2.1. Aitken nuclei mixing ratios and concentrations at various locations

Altitude [km]	South Polar (McMurdo) [No./mg]	No./cm^3	Equatorial (Panama) [No./mg]	No./cm^3	Midlatitude (Wyoming) [No./mg]	No./cm^3	North Polar (Fairbanks) [No./mg]	No./cm^3
10	200	83	6000	2481	1000	414	300	124
15	70	14	2500	487	300	58	70	14
20	35	3	50	4	50	4	30	3
25	10	0.4	40	2	40	2	-	-
30	-	-	30	0.6	30	0.6	-	-

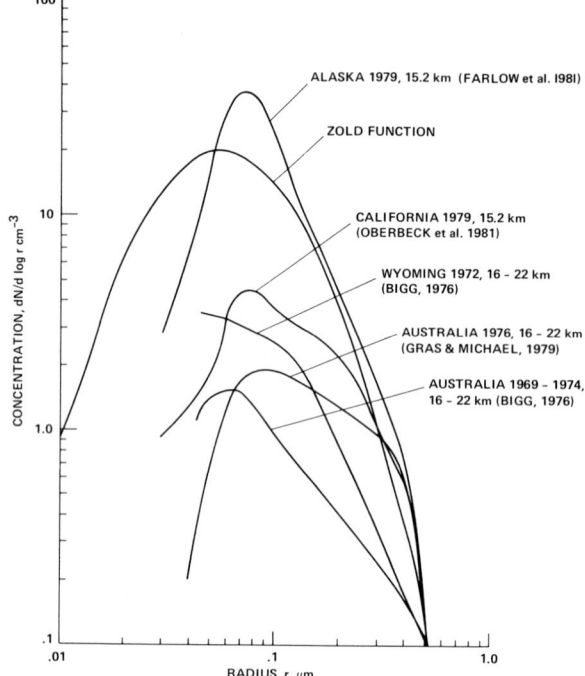

Fig.2.4. Average and individual size distributions at differten locations and times compared to a model zero-order logarithmic distribution function

distributions, concentrations, or compositions. This suggests that considerable variation exists in these properties worldwide. Figure 2.4 shows a collection of recent individual and averaged distributions [2.46,51,64,65]. These size-frequency curves are presented with an analytical average size distribution (zero-order logarithmic distribution — ZOLD-function) proposed earlier by TOON and POLLACK [2.59]. The distributions tend to be approximately monomodal, peaking in size near 0.1-μm radius or slightly below. It is noteworthy that concentrations differ by as much as a factor of thirty.

Two influences may account for these differing results. The first, an environmental one, is the change in stratospheric particle distributions that results when exploding volcanos inject large amounts of sulfur gases and ash. The second, an instrumentation problem, is the effect that ultrasteep size distributions have on required accuracies of the sampling devices [2.65]. In the first case, large eruptions initially alter aerosols in the region near the volcano, then cause changes throughout the hemisphere, and finally globally, if intrusion of volcanic debris is large. But effects tend to be larger in the volcano's region, proportionately smaller at remote sites. Hence real differences in measured values are likely at different sampling locations. These differences, however, tend to diminish in quiescent periods of little volcanic activity. In the second case, the precipitous drop in numbers of particles beyond the mode radius (Fig.2.4) as particle sizes increase dictates almost unattainable accuracy among various measuring systems to attain comparable values. As an example, an increase of 10 percent in particle radius in size distributions results in a 30 to 50 percent decrease in concentrations. Thus if accuracy of measuring particle sizes is ± 10 percent for each of two different sampling systems, their measured radii might differ by as much as 20 percent. Concentrations or other calculated properties based on particle size (such as aerosol mass) could then be different by 60 to 100 percent. Consequently, observed differences between results of various experimenters may not really be significant if cumulative errors in radii measurements of ± 10 to 20 percent are possible.

Compositional differences found in larger acid drops over the past two decades usually are small except following large volcanic ash intrusions. Most analyses support the conclusion that particles are impure sulfuric acid. Detection of $(NH_4)_2SO_4$ and $(NH_4)_2S_2O_8$ in earlier years [2.25,26,29,66,67] is questioned by HAYES et al. [2.61]. They found no crystals of $(NH_4)_2SO_4$ or $(NH_4)_2S_2O_8$ when they protected stratospheric particles from air contamination in sealed containers by an argon atmosphere. This prevents any particle reactions from the time of collection through insertion into the electron microscope. But when samples are exposed briefly to laboratory air, particles are quickly converted to ammonium sulfates by traces of ammonia gas. They confirmed the acidic composition of uncontaminated particles by identifying electron diffraction halos of sulfuric acid.

When volcanos inject volcanic products to the stratosphere, as in explosions of Volcán de Fuego, Guatemala, in 1974, and Mt. St. Helens, Washington, 1980, major aerosol changes occur. The most significant alteration at first is the addition of solid ash grains (perhaps coated with liquid acid) and sulfurous gases. Theoreticians believe acid builds up as a consequence of gas-to-particle conversion, accelerated or delayed by the amount of water vapor present.

LAZRUS et al. [2.68] noted such an acid buildup after the Fuego eruption. Worldwide background measurements before the eruption showed sulfate mixing ratios of 0.1 ppbm at the tropopause rising to 0.5 ppbm in the upper part of the layer near 20 km. Six months after the Fuego eruption the worldwide values had climbed to 0.4 ppbm at the tropopause to as much as 3.6 ppbm at the top of the layer. A year after this, values had declined at the higher altitude to about 1.6 ppbm, but remained the same at the tropopause. By 1979, sulfate concentrations throughout the layer remained in the 0.4 to 0.8 ppbm range [2.65,69].

In the Mt. St. Helens eruptions, however, some evidence suggests that either the explosions injected liquid acid, or the fluid formed very quickly, contrary to the slow buildup after Fuego [2.70]. Large particles of ash (up to 30 μm after the first eruption, 3 μm from later ones), encased with heavy acid coatings, fell quickly from the stratosphere [2.70]. After about two months and several circuits around the Earth, all ash and much acid was gone. Some effects of the Mt. St. Helens eruptions lingered, however. Six months after the initial eruption, scientists still observed a concentration of larger acid droplets about three times preeruption levels [2.71,72]. Whether size distributions of these residual droplets had changed is still being investigated, but their compositional makeup no longer included granular inclusions of ash.

2.2.8 Granular Inclusions

Only a few investigators have tried to isolate and examine granular inclusions in acid droplets. MOSSOP [2.73] isolated 3579 of these insoluble particles from seven stratospheric collections obtained during a nonvolcanic period but was unable to identify their composition with electron diffraction methods. For the few that were spherical he suggested an extraterrestrial origin. The others, he speculated, might have been volcanic ash although he knew of no recent eruptions. Or, they might have come from nuclear tests conducted the year before. FARLOW et al. [2.29], using a different method, also examined some undissolved granules collected in acid droplets. Granules obtained about a year after eruption of Volcán de Fuego contained elements suggestive of volcanic silicates. But collections made five months after that had only a few inclusions, without any silicates. These investigators, now examining acid droplet collections made several months after the Mt. St. Helens eruptions, see virtually no residual granules. The presence of such materials in acid droplets can alter the way in which they absorb and scatter radiation, so the search is significant for theorists developing energy balance models for climate impact studies.

2.2.9 Cosmic Dust

BROWNLEE [2.74], and GANAPATHY and BROWNLEE [2.75] conclusively identified interplanetary dust particles obtained by collectors on U-2 aircraft. They used neutron activation and energy dispersive X-ray methods to determine the particles as carbonaceous chondrites. Their methods culminated a two-decade search for ways to uniquely identify extraterrestrial particles retrieved from the atmosphere. Their pioneering work now opens the way for elaborate studies of these rare constituents [2.76]. An interesting contaminant usually collected in the same particle size group as micrometeorites is the product of solid-fuel rocket exhausts — aluminum oxide spherules [2.77]. Apparently these spheres of about 3 µm diameter and larger are spread throughout the world by upper atmospheric winds. Researchers using U-2 aircraft have measured inside exhaust plumes of rockets soon after they were launched [2.27]. They collected spherules identical to those other investigators found in the free stratosphere, confirming their origin. In nonvolcanic periods, these contaminating spherules comprise the major group of particles in the 3 to 8 µm diameter range [2.77].

2.2.10 Volcanic Ash

After the eruption of Mt. Agung, Bali, in 1963, MOSSOP [2.78] collected acid-coated ash samples with U-2 aircraft. He found ash particle sizes decreased from a range of about 0.2 to 8 µm diameter in the first month after the eruption to a range of 0.1 to 1 µm a year later. Significantly, many ash grains from Agung remained in the stratosphere even after a year, in contrast to Volcán de Fuego and Mt. St. Helens where ash quickly disappeared.

FARLOW et al. [2.79] in a study of Mt. St. Helens ash grains, found most size distributions varied from sample to sample (Fig.2.5). An examination of size distributions shows that high-altitude specimens generally have particle size modes around 0.3 µm radius, while lower altitude samples have modes at about 0.6 µm. Even though distributions coincide at the large-particle end, there is a marked decrease in the small-particle component at lower altitudes. Sizes of ash grains from the Mt. St. Helens eruptions are not much different from those MOSSOP [2.78] observed after the Mt. Agung explosion. Nevertheless, grains from Mt. St. Helens were so heavily encased in acid that these large globules quickly fell from the stratosphere.

FARLOW et al. [2.79] derived ash grain mineralogy for each of the Mt. St. Helens samples using nondispersive X-ray analyses that provided elemental compositions. Although there were no significant variations of composition with particle size, proportions of major minerals varied widely between samples. Glasses ranged from 40 percent to 100 percent, while plagioclase varied from almost none to a high of 32 percent. They found amounts of hornblende and pyroxene varying from a trace to 22 percent. This diversity of composition for samples from different parts of the cloud showed how inhomogeneous it was.

2.2.11 Variations

Aircraft and balloon measurements of aerosol properties have established their vertical and geographical structure and temporal trends. Because the tropopause is lower at the poles, the aerosol layer projects down to about 10 km in polar latitudes, but forms only above about 15 km at the equator [2.68]. Even when the stra-

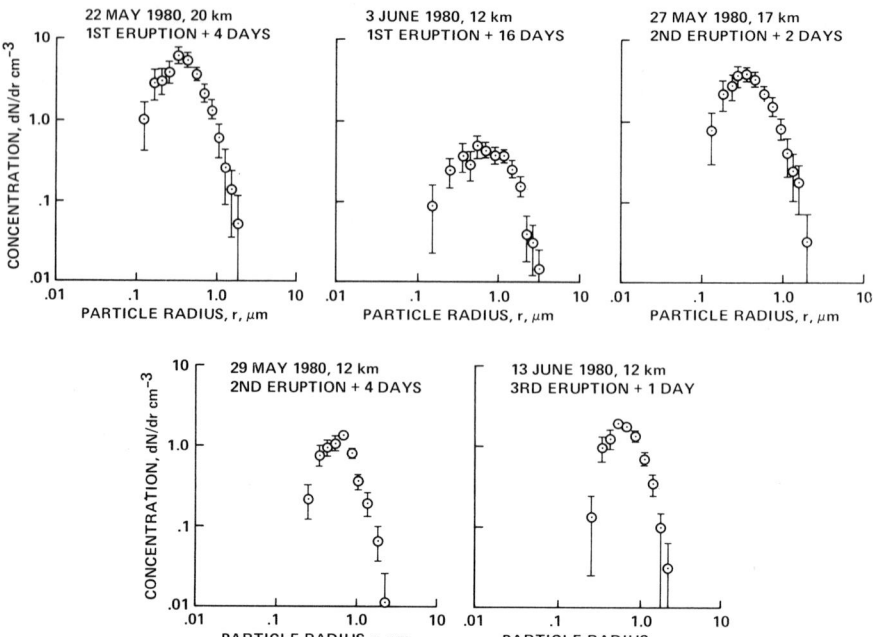

Fig.2.5. Ash grain particle size distributions in the stratosphere for three Mt. St. Helens eruptions, May and June 1980. Error bars are 95% confidence intervals for populations as inferred from samples

tosphere is perturbed by volcanic eruptions, low temperatures at the tropopause control lower fringes of the layer.

Seasonal variations in tropopause height, however, lead to seasonal fluctuations of total aerosol concentrations in a vertical column through the stratosphere [2.80]. The summer aerosol layer is very stable, with the lower stratosphere near the tropopause remaining relatively free of particles. In winter, particle concentrations in the lower stratosphere seem to increase, often having a layered structure. Spring conditions bring on a variable, highly layered structure in both lower stratosphere and upper troposphere. Prior to spring conditions the upper troposphere is generally free of particles. Aerosol conditions in the layer above 20 km, however, are quite stable throughout the seasons [2.64,80]. LAZRUS and GANDRUD [2.81], on the other hand, noted significant seasonal changes in sulfate mixing ratios above 20 km. But they collected all sizes of sulfate particles with their efficient filter sampling method, whereas HOFMANN et al. [2.82] detected only particles larger than 0.3 μm diameter with their light-scattering device. This suggests that small particles above 20 km may fluctuate seasonally. More research is needed to resolve this question.

Periodic volcanic eruptions over the past decade have influenced long-term aerosol trends the most. The explosion of Volcán de Fuego in 1974, for example, caused

Table 2.2. Larger acid droplets (> 0.3 μm diam) peak mixing ratios and concentrations at midlatitudes

Year	Peak Mixing Ratios [No./mg]	Peak Concentrations [No./cm^3]
1971	20	2
1972	8	1
1973	6	0.7
1974	80	10
1975	20	2
1976	15	1.8
1977	8	1
1978	7	0.8
1979	6	0.7
1980 (Sierra Negra)	12	1.4
1980 (Mt.St.Helens)	30	4

major changes in stratospheric aerosol loading. HOFMANN and ROSEN [2.83] observed these changes in maximum particle concentration for particles larger than 0.3-μm diameter. Table 2.2 generalizes their findings in the period 1971 to 1979 to show a decreasing particle concentration up to mid-1974 when a dramatic rise occurred due to the Fuego eruption, followed by gradual decline to a fairly constant level in early 1979. This background is then perturbed by volcanic input from La Soufrière on the island of St. Vincent in the Caribbean and from Sierra Negra in the Galapagos Islands [2.56]. Mt. St. Helens injected huge amounts of volcanic products in May 1980, which overwhelmed this small intrusion in the northern hemisphere. After Mt. St. Helens ash settled from the stratosphere, remaining aerosol background was several times higher than preeruption values [2.70-72].

2.2.12 Discussion

Additional sensors using light-scattering methods are now in use in the stratosphere on balloons and aircraft. GRAS and MICHAEL [2.46] described size distributions obtained from their balloon-borne photoelectric particle counter that agree with collections made earlier. This photoelectric counter provides complete size distributions for all particles larger than 0.3 μm diameter, contrasting with other systems that detect size groups [2.84].

Knollenberg (personal communication, 1981) has installed a photoelectric particle counter on U-2 aircraft for stratospheric measurements of particles larger than 0.1 μm diameter. His system also provides complete distributions to about 3 μm. MIRANDA et al. [2.85], and MIRANDA and DULCHINOS [2.47] used laser illumination of particles to obtain complete size distributions in the stratosphere on several balloon flights. Use of these newer sensors should provide additional measurements of stratospheric aerosols in the future.

Applying microbeam analytical techniques, such as SIMS and ESCA, and Raman spectra microanalysis, should increase knowledge about aerosol surfaces and molecular composition. Although established analytical methods like chemical analysis, electron diffraction, and X-ray evaluation are essential, additional capabilities of microbeam systems add new dimensions to particle analyses. In particular, investigators need to identify noncrystalline compounds and trace constituents.

Instrumentation errors affecting particle size determinations need to be reduced if results are to be effectively compared. This is because large inaccuracies in calculated parameters result from small measurement errors of particle sizes. Contributing to these errors are effects due to nonisokinetic sampling at instrument intake probes, instrument collection or detection inefficiencies for particles of different sizes, and contamination of samples in uncontrolled environments.

Scientists have learned a lot in the past two decades about stratospheric aerosols using balloon and aircraft sampling methods. They know a moderate amount about average physical, chemical, and optical properties of particles. They perceive to some extent how particles form, grow, and disperse worldwide. They know somewhat the distribution of particles vertically and geographically in volcanic and quiescent times. But they still seek elusive details needed to understand all aspects of the processes. They especially need measurements that better define particle formation and growth before and after volcanic explosions. Identification of subtle sources and sinks that supply and deplete aerosol layers in quiet times are necessary. And what are the transport mechanisms that carry particles across meterological boundaries? How do optical properties of particles change as trace constituents vary? What are all the chemical constituents within the particle matrix? These, then, are some of the unknowns that continue to pique scientific curiosity about this complex atmospheric system.

2.3 Lidar Measurements

Stratospheric aerosols and the lidar measurement technique have a rather long association. The stratospheric aerosol layer was one of the first atmospheric constituents to be measured by the emerging lidar technique, beginning in 1963. Since then, lidar measurements have contributed a substantial portion of the existing data set on stratospheric aerosol behavior. Lidar measurements in North America, South America, Europe, Asia, Australia, Japan, Hawaii, and Greenland have documented the stratospheric aerosol's response to several major volcanic injections and have helped to determine the vertical and latitudinal structure of the quiescent, or background, layer. Comparisons of lidar measurements with direct sampling by aircraft and balloons have helped to confirm aerosol optical models while leading to a more thorough understanding of each measurement technique involved (including lidar itself). Most recently, lidar measurements have helped to validate the performance of satellite sensors of the stratospheric aerosol.

This section summarizes the principles of the lidar measurement technique and then reviews selected examples of the measurement data set and their application to stratospheric aerosol studies.

2.3.1 The Lidar Technique

Lidar is an acronym for light detection and ranging. The technique is analogous to radar (radio detection and ranging), but uses light (strictly speaking, ultraviolet through infrared radiation) as the probing energy, rather than radio waves. A limited number of lidar measurements were performed before the invention of the laser, using searchlights as sources. However, the advent of the laser provided a far superior source, and caused a rapid growth in the range of measurements possible with the lidar technique. Today all lidars use lasers as sources, and the terms "lidar" and "laser radar" have become synonymous.

a) *Physical Principles and Equipment*

The lidar technique works by firing a laser pulse into the atmosphere and measuring, as a function of time after pulse transmission (hence, range from the lidar), the radiation scattered back to the lidar by the gases and particles that comprise the atmosphere. The broad range of measurements possible with the lidar technique has recently been reviewed in the book by HINKLEY [2.86].

Figure 2.6 shows a simplified block diagram of a typical elastic-backscatter lidar (i.e., one that measures backscattered radiation at the transmitted wavelength λ). Stratospheric aerosol lidars to date have used either ruby ($\lambda = 0.694$ μm), dye ($\lambda \approx 0.45$ to 0.6 μm) or Nd:YAG ($\lambda = 1.06, 0.53$ μm) lasers. The receiver telescope primary mirror or lens is typically at least 0.3 m in diameter, to collect a sufficient amount of the relatively weak stratospheric backscatter signal. After signal conditioning (filtering, amplification, etc.), the detector output is recorded, typically using a computer-based data system, which allows for initial data processing and display. More detailed descriptions of ground-based stratospheric aerosol lidars are given by McCORMICK and FULLER [2.87], RUSSELL et al. [2.88], REITER et al. [2.89], and HIRONO et al. [2.90], among others.

Stratospheric aerosol lidars have also been operated on aircraft. Airborne operations offer the following advantages:
— The airborne platform can be operated above low cloud cover and other adverse weather that would curtail ground operations;
— the airborne lidar can readily be moved to locations of special interest;
— the flight path can be chosen so that the lidar data describe a geometric plane of special interest.

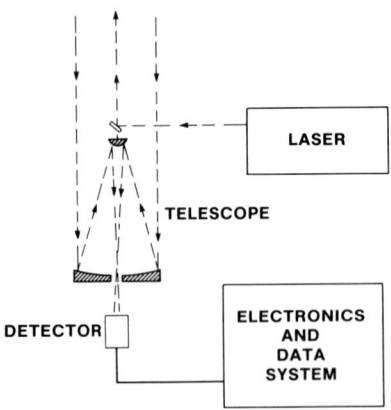

Fig.2.6. Schematic diagram of a typical stratospheric aerosol lidar

These advantages are especially useful in documenting large-scale aerosol structure, mapping volcanic injections, and in coordinated, multisensor experiments, including those conducted to validate satellite sensors. Airborne stratospheric lidars are described by FOX et al. [2.91], FERNALD et al. [2.92], and FULLER et al. [2.93].

The prospect of making space-borne lidar observations of many constituents, including stratospheric aerosols, is an attractive one that has been studied by several groups and investigators in recent years [e.g., 2.94,95]. This possibility is discussed in more detail in Sect.2.4.

b) *Analysis Techniques*

For a vertically pointing lidar operating at wavelength λ, the detector output signal caused by scatterers at height z is given by

$$S(\lambda,z) = \frac{K(\lambda)}{(z - z_L)^2} Q^2(\lambda,z_L,z)[B_p(\lambda,z) + B_g(\lambda,z)] \quad , \tag{2.1}$$

where $K(\lambda)$ is a calibration factor (including transmitted energy, receiver area, and system optical and detection efficiencies), z_L is the lidar height, $Q^2(\lambda,z,z_L)$ is the two-way transmission between z and z_L, and B_p and B_g are particulate and gaseous backscattering, respectively. Stratospheric aerosol lidar data are nearly always analyzed in terms of the scattering ratio, defined as

$$R(\lambda,z) \equiv \frac{B_g(\lambda,z) + B_p(\lambda,z)}{B_g(\lambda,z)} = 1 + \frac{B_p(\lambda,z)}{B_g(\lambda,z)} \quad . \tag{2.2}$$

Combining (2.1) and (2.2) shows that $R(\lambda,z)$ can be derived from the lidar signal as

$$R(\lambda,z) = \frac{(z - z_L)^2 S(\lambda,z)}{K(\lambda)Q^2(\lambda,z,z_L)B_g(\lambda,z)} \quad . \tag{2.3}$$

In solving this equation, $S(\lambda,z)$ is provided by the lidar measurement, $K(\lambda)$ is determined through a normalization procedure, $Q^2(\lambda,z,z_L)$ is usually provided by a model (which may be updated using lidar data), and $B_g(\lambda,z)$ is determined from a current radiosonde- or satellite-inferred density profile, or from a model. The normalization procedure for determining $K(\lambda)$ in effect uses the atmosphere itself, at the height z^* where $R(\lambda,z)$ attains its minimum value, as a calibration target of (approximately) known backscatter coefficient. The probable errors in this procedure and in the other inputs to (2.3) have recently been discussed in the literature [2.96] (see also below).

Given the scattering ratio values obtained from (2.3), one can obtain the absolute backscattering coefficient by rearranging (2.2), i.e.,

$$B_p(\lambda,z) = [R(\lambda,z) - 1]B_g(\lambda,z) \quad . \tag{2.4}$$

Figure 2.7 shows typical profiles of scattering ratio and particulate backscattering coefficient derived from a nonvolcanic lidar measurement using (2.3,4). The error bars in Fig.2.7 were obtained by the procedure described in [2.96]; they include uncertainties in the measured lidar signal S, the calibration factor K, the transmission Q^2, and the gas backscattering coefficient B_g (hence, total gas density).

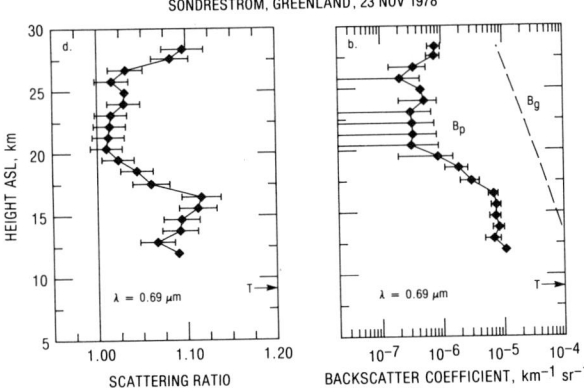

Fig.2.7. Vertical profiles of scattering ratio and particle backscattering coefficient derived from lidar measurements

2.3.2 Survey of Results

a) *The Post-Agung Decay Period (1963-1972)*

The volcano Agung in Indonesia (115°E, 8°S) exploded in early 1963, injecting very large amounts of particles and gases into the stratosphere. By coincidence, the first laser radar observations of the atmosphere were made during this year in Massachusetts [2.97,98] and California [2.99]. The Massachusetts observations of GRAMS and FIOCCO [2.100] focused on the stratosphere and documented the behavior of the stratospheric aerosol layer throughout 1964 and into 1965. Their results are summarized in Fig.2.8 in terms of the profile-maximum value of the backscatter

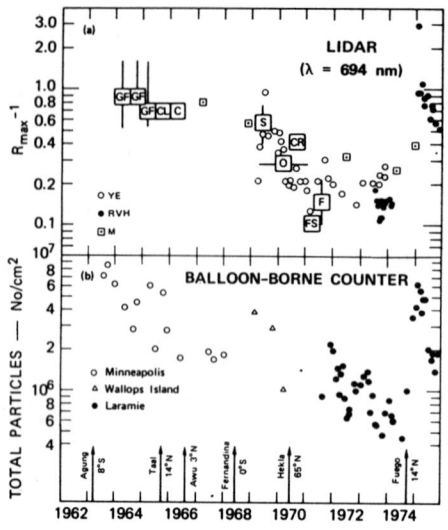

Fig.2.8. Comparison of lidar and balloon-borne particle counter (dustsonde) measurements of the stratospheric aerosol, 1963-1975. (a) Maximum ratio of particulate to gaseous backscattering measured by a number of stratospheric lidar groups. Boxes indicate mean of many observations, bars indicate range in time and magnitude. Locations and references: (GF) Massachusetts [2.100]; (CL) California [2.101]; (C) Jamaica [2.102]; (M) Virginia [2.103]; (RVH) California [2,88,104]; (YE) Australia [2.105]; (CR) Brazil [2.106]; (S) Colorado [2.107]; (O) Jamaica [2.108]; (FS) Colorado unpublished ; (F) Hawaii and Bermuda [2.91].
(b) Number of particles (radius ≥ 0.15 μm) above tropopause as measured by dustsonde [2.48,80]. Arrows give times of volcanic eruptions thought to have penetrated the tropopause significantly

mixing ratio, R-1 [i.e., B_p/B_g; cf.(2.2)]. As can be seen, the 1964-1965 values are unusually large in comparison to northern hemispheric values observed for the following eight years. These large values reflect the presence of the 1963 Agung effluent, which was transported from its southern hemispheric injection point into the northern hemisphere, where it influenced stratospheric aerosol concentrations for at least several years. The Agung volcanic effects were also documented by dustsonde measurements (Sect.2.2), which are shown in Fig.2.8b for comparison to the lidar data.

Subsequent lidar measurements of the post-Agung stratospheric aerosol were made by COLLIS and LIGDA [2.101] in California, CLEMESHA et al. [2.102] in Jamaica, and McCORMICK [2.103] in Virginia; their results are also shown in Fig.2.8. Notice that, during 1965-1968, both the lidar and the dustsonde results show that the decrease in aerosol concentration was very slow or even negligible. As suggested by the arrows at the bottom of Fig.2.8, this interruption of the 1964-1965 decay was probably caused by the new volcanic injections of Taal, Awu, and Fernandina. However, there is another factor that no doubt also contributed to the slowness of decay. Notice that the 1963-1965 decay (i.e., before the Taal eruption) was also considerably slower than that observed at northern midlatitudes after the 1974 eruption of Fuego (a northern hemisphere volcano). One reason for the slow northern hemisphere decay of Agung effects is that Agung is located in the southern hemisphere, where it caused stratospheric aerosol loadings a factor of 10 or more larger than those observed in the northern hemisphere [2.109]. These large southern hemisphere concentrations served as a continuing source to the northern hemisphere for at least several years.

Beginning in 1969, the number of active stratospheric lidars increased considerably, as reflected by the increased number of data points in Fig.2.8. In the period

1969-1972, the first southern hemispheric measurements were made by GAMBLING et al. [2.105] in Australia and by CLEMESHA and RODRIGUES [2.106] in Brazil. Northern hemispheric measurements were made by SCHUSTER [2.107] and FRUSH and SCHUSTER in Colorado, and by OTTWAY [2.108] in Jamaica. In 1971 FOX et al. [2.91] made the first airborne lidar observations of the stratospheric aerosol. Over the 1969-1972 period both lidar and dustsonde measurements recorded a considerable decrease in stratospheric aerosol concentrations, as volcanic effects were gradually removed and background concentrations were approached.

b) *The CIAP Background Period (1972-1974)*

The period 1972-1974 was a time of considerable stratospheric measurement activity, conducted as a part of, or adjunct to, the Climatic Impact Assessment Program (CIAP). This program resulted from widespread concern that proposed fleets of stratospheric aircraft could decrease the stratospheric ozone layer and adversely affect life on the Earth's surface. As a part of CIAP, ground-based lidar measurements were made by NASA Langley Research Center [2.110] and SRI International [2.88,104], and airborne measurements were made by the National Center for Atmospheric Research [2.111]. The NASA Langley results are described below. The SRI results are shown in Fig.2.8, which documents the background or near-background concentrations observed by both balloon and lidar in 1974.

Figure 2.9 shows a meridional cross section of scattering ratio derived from the NCAR airborne measurements of FERNALD and SCHUSTER [2.111]. The latitudinal variation of the peak altitude (dashed line) is similar to that revealed by the February-March and April 1973 dustsonde measurements reported by ROSEN et al. [2.84] (see also Sect.2.2). However, as noted by FERNALD and SCHUSTER, the peak shown in Fig. 2.9 at latitudes north of $70°N$ is at an unusually high altitude because of an upper-tropospheric low-pressure region present during the single high-latitude flight; also, absolute scattering ratio values between the equator and $30°N$ are likely to be too small because of profile normalization between 28 and 30 km, a layer that contains significant aerosol mixing ratios at those latitudes (see the discussion of errors in Sect.2.3.1).

c) *The Fuego Injection and Decay (1974-1977)*

The volcano Fuego in Guatemala ($14.5°N$, $91°W$) erupted violently in October 1974, greatly increasing northern hemispheric stratospheric aerosol concentrations. These effects were measured by a variety of techniques, including the ground-based lidars of McCORMICK and FULLER [2.112], REMSBERG and NORTHAM [2.113], FEGLEY and ELLIS [2.114], FERNALD and FRUSH [2.115], HIRONO et al. [2.116], CLEMESHA and SIMONICH [2.117], and RUSSELL and HAKE [2.104]. The results of RUSSELL and HAKE are summarized in Fig.2.8; note the increase by a factor of 10 or more, and the subsequent decay, which parallels that of the dustsonde data.

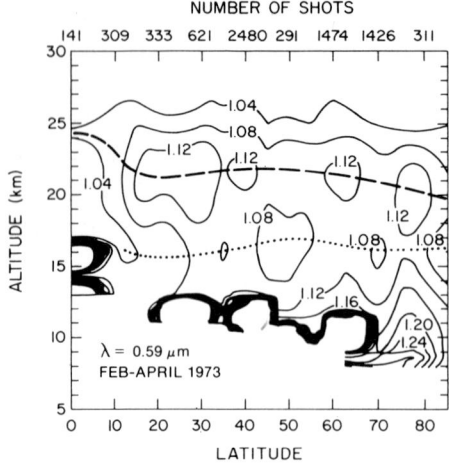

Fig.2.9. Meridional aerosol scattering ratio cross section ($\lambda = 0.59$ µm) produced from a March-April 1973 set of airborne lidar measurements. Heavy, close-packed contours at bottom mark circus cloud tops [2.111]

McCORMICK et al. [2.118] reported a detailed series of 34 pre- and post-Fuego measurements that spanned nearly two years. Their vertical profiles of scattering ratio are shown in Fig.2.10, along with the simultaneous radiosonde temperature profiles. Notice the initially very complicated structure, with multiple layers that changed markedly from one observation to the next (intervals of ~ 1 wk). Maximum aerosol layer backscattering was not attained until three or four months after the October injection, as a consequence of both the northward transport required (from 14.5 to 37.1°N) and the time for particle formation and growth from the injected gases. The very large scattering ratios attained in January 1975 (ranging up to 4.2) decayed thereafter, approaching the pre-Fuego values of ~ 1.1 by the end of 1976. At the same time, the initial multiple layers merged to form a single, broader peak with a centroid that rose slowly toward the pre-Fuego height of 21 to 22 km (see also [2.104]).

Figure 2.11 [2.119] summarizes the peak backscatter mixing ratios $R_{max}-1$ of McCORMICK et al. and extends the series through 1980. (The dustsonde-inferred curve in this figure is discussed in Sect.2.3.3). This time series yields a background-corrected 1/e decay time of 8 months over the January 1975 - July 1976 period [2.119,120]. This value agrees fairly well with the value of 10 months obtained from the February-November 1975 lidar data of RUSSELL and HAKE [2.104] when a background correction is applied [2.120].

Southern hemispheric effects of the Fuego injection were documented by the Brazil lidar observations of CLEMESHA and SIMONICH [2.117]. Among other things, these observations, made at 23°S latitude, showed a six-month delay between the Fuego eruption (October 1974) and the arrival at the lidar site of noticeable stratospheric effects (April 1975). This delay was attributed to the inhibition of southward eddy transport during the southern summer by the mean meridional circulation.

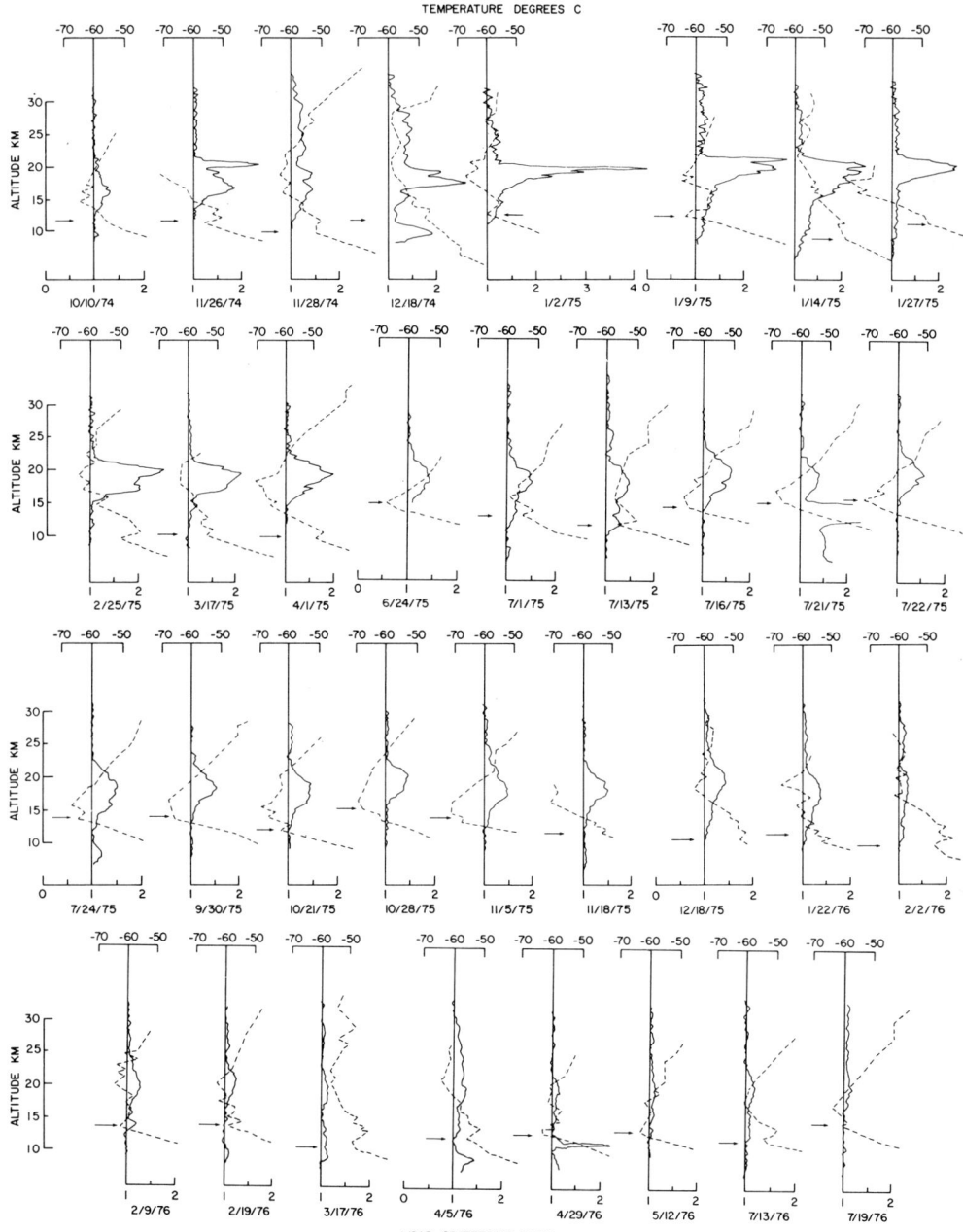

Fig.2.10. Pre- and post-Fuego stratospheric aerosol scattering ratio profiles ($\lambda = 0.69$ μm; solid lines) and temperature profiles (dashed line) measured in Virginia, October 1974–July 1976. [2.118]

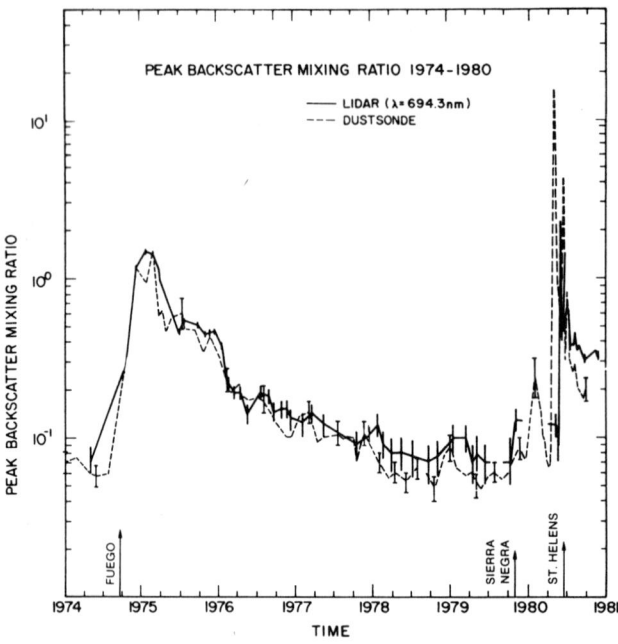

Fig.2.11. Comparison of a seven-year series of peak stratospheric backscatter mixing ratio, $R_{max}-1$, measured by lidar (solid line) with that inferred from dustsonde and optical models (dashed line). [2.119]

CLEMESHA and SIMONICH 2.117 also showed time series of vertical profiles and of peak and column backscattering (analogous to Figs.2.10,22) that include the pre-Fuego background period (1972-1974), the Fuego increase (1975), and the decay period extending into 1977.

d) *The Pre-St. Helens Background Period (1977-1980)*

The period extending roughly from mid-1977 through early 1980 has been cited by several studies using several measurement techniques as a background period when the stratospheric aerosol was free from major volcanically induced increases. This is evident, for example, in the northern midlatitude dustsonde data of HOFMANN and ROSEN [2.120] (see Sect.2.2), in the northern midlatitude lidar data of REITER et al. [2.89], and in those shown in Fig.2.11. However, the absence of effects that are readily identifiable for long periods of time as having a specific volcanic source does not rule out the influence of less explosive or less massive volcanic eruptions which, taken together, may be a primary sustainer of the background stratospheric aerosol layer. In fact, during this so-called background period, several minor volcanic effects were noted by lidar, in addition to the transient increases noted by dustsonde [2.56] (Sect.2.2.11). For example, FEGLEY et al. [2.121] reported an upper tropospheric layer, thought to have originated at the volcano Nyiragonga (1.5°S, 29.2°E), that appeared to penetrate the tropical tropopause in January 1977, possibly because of convection induced by radiative heating in the layer. Also, the NASA Langley airborne stratospheric lidar documented stratospheric penetrations by the Soufrière volcano (13.3°N, 61.2°W) in February while making satellite validation measurements [2.122]. Previously, transient effects attributed to the St. Augustine volcano (59°N, 135°W) had been reported by REMSBERG [2.123].

e) *The St. Helens Injection and Decay (1980)*

The Mt. St. Helens eruptions of May 1980 are unquestionably the best-documented example of volcanic influence on the stratosphere, and lidar measurements have played no small role in this documentation, especially of global-scale transport. The arrival time, height, magnitude, and duration of volcanic effluent have been documented by ground-based lidars in California, New Mexico, Colorado, Illinois, Wisconsin, Virginia, England, France, Germany, Italy, and Japan, and especially by the NASA Langley airborne stratospheric lidar, operating at many locations in the U.S. and over the North Atlantic ocean [2.124]. McCORMICK has assembled a number of the ground-based observations to develop a consistent picture of the zonal transport of the eruption clouds at various heights [2.125]. An example is shown in Fig.2.12, which applies to layers above 20 km in altitude. The shaded area indicates the expected range of volcanic cloud longitudes as a function of time, based on a zonal wind model [2.126]. The points show the actual arrival times noted by lidar measurements at nine sites in Asia, Europe, and North America, as well as data from dustsonde flights and the SAGE satellite. Note that the transport above 20 km is to the west; hence, the North American observations result from a nearly complete global circuit. McCORMICK [2.125] also derived from the lidar data a volcanic increase in stratospheric aerosol mass loading of 0.4 to 0.6×10^6 tonnes, a value that agrees well with inferences from the SAGE satellite data.

A typical example of a post-St. Helens time series of vertical profiles is shown in Fig.2.13. These data, measured in Germany by REITER et al. [2.127] are similar to other European and North American observations in showing highly variable vertical structure, the early arrival of material below 20 km (resulting from eastward winds), and the later arrival of material above 20 km (resulting from westward winds).

2.3.3 Comparisons of Lidar and Other Results

a) *Dustsondes*

Dustsondes (Sect.2.2) provide the directly measured stratospheric aerosol data that are most readily compared to lidar data, because they yield a profile with fine vertical resolution and describe integral properties (numbers of particles larger than two size cutoffs) that are related to the volume backscatter coefficient. As a result, many comparisons between lidar and dustsonde measurements have been made. The first comparisons made with a collocated lidar and dustsonde were

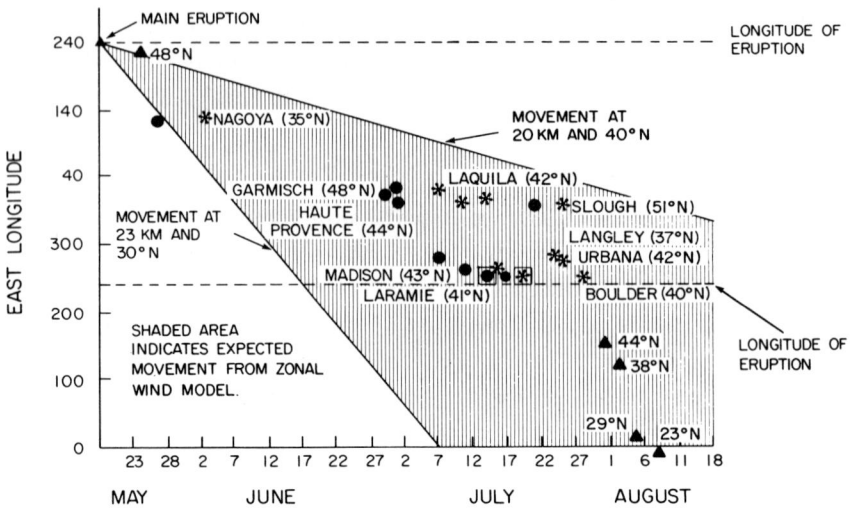

Fig.2.12. Arrival times of post-St. Helens aerosol layers above 20 km as observed by various lidar stations around the globe. * 20-22 km layer; ● 22-24 km layer. Crosshatched area shows range of stratospheric air movements expected from a model [2.126]. Also shown are dates and longitudes of SAGE measurements (▲) and of dust-sonde measurements (boxed points) at Laramie, Wyoming. [2.124]

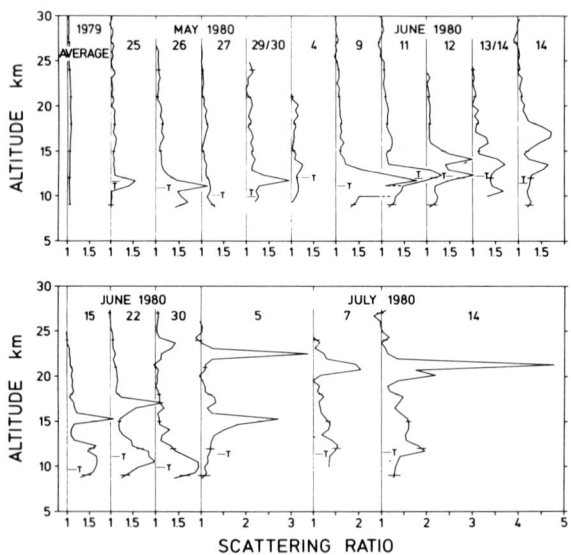

Fig.2.13. Post-St. Helens scattering ratio profiles measured by ruby lidar ($\lambda = 0.694$ μm) in Germany). T denotes Munich radiosonde tropopause. [2.127]

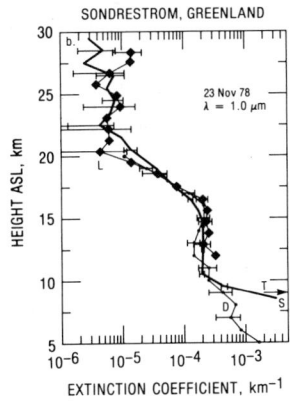

Fig.2.14. Comparison between stratospheric particulate extinction profiles ($\lambda = 1.0$ μm) measured by satellite (S) and inferred from dustsonde (D) and lidar (L) measurements [2.128]

conducted in 1972 and described by NORTHAM et al. [2.110]. Since then, comparisons of collocated lidar and dustsonde measurements have been made in Wyoming, Greenland, Alaska, Brazil, and Texas as part of the SAM II and SAGE satellite validation programs. A typical example from Greenland is shown in Fig.2.14, from [2.128]. In this case both the lidar (backscatter) and the dustsonde (number) data have been converted to best-estimate extinction profiles using an optical modeling technique [2.129]. This technique uses aerosol size distributions and refractive indices that are constrained to be consistent with the actual dustsonde number measurements and with previous composition data. However, this constraint still allows a range of optical models, and hence of conversion factors. The uncertainty in conversion factors combines with the lidar and dustsonde measurement uncertainties to yield the error bars shown in Fig.2.14. As in most of the lidar-dustsonde comparisons made to date, the differences between the two results shown are consistent with these uncertainties.

A seven-year series of comparisons, in which the lidar ($37.1°N$, $76.3°W$) and dustsonde ($41.2°N$, $105°W$) were located at nearly the same latitude but separated by ~ 3000 km in longitude was conducted by SWISSLER et al. [2.119]. In that study the same modeling technique described above was used to convert the two-channel dustsonde data to expected values of peak backscatter mixing ratio. The results are shown in Fig.2.11 as a dashed line. As can be seen, the overall agreement is quite good; however, in the 1978-1979 background period, the model used in Fig. 2.11 yields expected backscatter mixing ratios that slightly but consistently underestimate the lidar measurements. The implications of this underestimation for stratospheric aerosol optical properties were discussed by SWISSLER et al. [2.19], who also used the optical modeling technique to show how the mean particle radius for backscatter changed in response to dustsonde-inferred size distribution changes after the Fuego eruption.

b) *Aircraft Sampling*

Lidar measurements have also been compared to mass concentrations and number-vs-size distributions measured by aircraft. A July 1973 experiment reported by RUSSELL et al. 2.88 yielded agreement between a filter mass measurement made at 19 km by a WB-57F aircraft and the value inferred from simultaneous ground-based lidar measurements by using a range of optical models. More extensive comparisons were made in the July 1979 experiments conducted in Alaska as part of the SAM II and SAGE satellite validation and Aerosol Climatic Effects (ACE) programs [2.130-132]. Those comparisons showed agreement between filter-measured and lidar-inferred mass at all heights between 12 and 18 km. Impactor-measured particle size distributions agreed (to within experimental uncertainties) with the filter and lidar results at 15 km, but implied considerably more mass and backscatter than the measured values at 18 and 20 km. These differences are now being analyzed as a guide to laboratory tests of the various sensors and further field experiments.

c) *Satellites*

Ground-based and airborne lidar measurements in North America, South America, Europe, Japan, and Greenland have played a major role in validating the data measured by the satellite sensors SAM II and SAGE (Sect.2.4). Figure 2.14 has shown an example of one lidar-satellite comparison, and others are discussed in Sect.2.4.

2.3.4 Additional Applications

In addition to the uses of lidar data mentioned thus far, two other applications are worthy of note. In the first, CADLE et al. [2.133,134] used lidar data to test a two-dimensional model of volcanic cloud dispersal and to infer the relative strengths of the Agung and Fuego injections (see also [2.116,117]). In the second RUSSELL and HAKE [2.110,135] combined lidar-inferred optical depths with radiative models to infer radiative and thermal consequences of the post-Fuego volcanic aerosol. Selected results are shown in Fig.2.15. They predict a Fuego-induced increase in the hemispheric-average Earth-plus-atmosphere albedo of about 0.007 or less, and stratospheric heating rates of about 4K or less per 100 days. At the surface they predict an equilibrium temperature decrease of about 1K or less; in fact, actual surface temperature decreases can be expected to be considerably less than the equilibrium value, because the Earth-surface thermal equilibration time ($\sim$ 4 years or more) is long compared to the Fuego stratospheric residence time ($\sim$ 8 months).

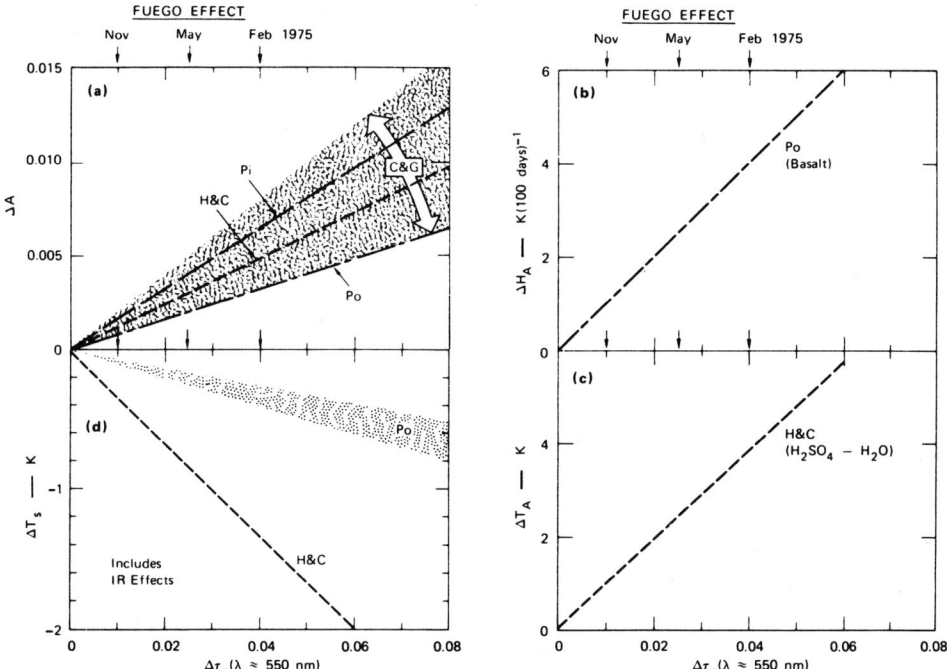

Fig.2.15a-d. Lidar-inferred optical depths of the 1975 post-Fuego volcanic stratospheric aerosol layer (arrows at top), with model predictions of changes in (a) earth-atmospherid albedo, (b) stratospheric heating rate, (c) equilibrium stratospheric temperature, and (d) equilibrium surface temperature. References for models are: C&G [2.136]; Po [2.104,137]; Pi [2.52]; H&C [2.138]. [2.135]

2.4 Satellite Observations

The conventional measurements of stratospheric aerosols described earlier in this chapter using such techniques as ground-based lidar or airborne particle counters and samplers, are limited in both spatial and temporal coverage. Obviously, a sensor developed for measuring aerosols from earth orbit allows a tremendous increase in the opportunities for measurements at different locations and times. These measurements have the unique advantage of being unaffected by weather conditions and provide a nearly continuous global coverage, which has not been available until now.

Figure 2.16 illustrates various techniques for the remote sensing of stratospheric aerosol properties from Earth-orbiting spacecraft. These techniques can generally be divided into two categories: active and passive. The passive method utilizes the sun or some other source such as the moon to provide a signal that may be attenuated and/or scattered by aerosol particles in the intervening atmosphere between the source and the spacecraft. The received signal is then inter-

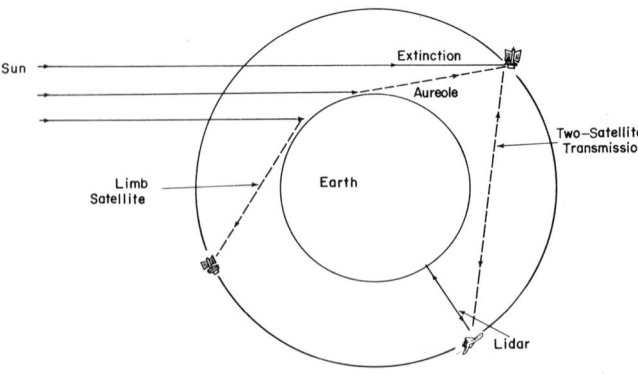

Fig.2.16. Diagram illustrating satellite observations of stratospheric aerosols using solar extinction, aureole, limb scattering, two-satellite transmission, and lidar techniques

preted in terms of aerosol extinction or scattering properties of the intervening aerosol. There are three measurement techniques associated with this method. The first technique measures extinction properties of the atmosphere during occultation of the light source as viewed from the satellite and is currently being used by two earth-orbiting spacecraft sensors [2.139] to be described later in this section. The second technique measures near-forward scattering properties of the aerosol. This is called the aureole technique [2.140]. Research efforts with this technique have mainly concentrated on ground-based observational systems which can only provide information on vertically integrated aerosol scattering properties [2.141]. The addition of an aureole measurement capability to the existing space-borne extinction instruments could potentially provide a much more complete description of stratospheric aerosol optical properties, for example, size distribution. The third technique is limb scattering, which measures the angular scattering properties of aerosols at different wavelengths and polarizations. This technique has the obvious advantage of much more global coverage per spacecraft orbit but suffers from a more difficult data inversion or interpretation process [2.142].

The active method consists of using artificial light sources such as lasers aboard satellites to actively probe the atmosphere. There are two measurement techniques associated with this method. The first uses a lidar system to measure the backscattering properties of stratospheric aerosols (Sect.2.3). A more detailed study of its potential for measuring aerosols and clouds alone has been carried out by RUSSELL et al. [2.95]. The second technique uses a light source, possibly a laser, on one satellite and a retroreflector or receiver on another. This technique is similar to the passive forward extinction technique but has the advantage of increased spatial and temporal coverage by properly tailoring the two satellite's orbital geometries. A variation of the technique is to have receivers or retroreflectors on the ground but because of the small stratospheric component to total extinction, it would probably not be very useful for stratospheric studies except during periods of large volcanic enhancements.

As stated earlier, the only technique that has been utilized in space to date is that of solar occultation or limb extinction. This section will describe in detail the two observational systems presently in earth orbit, which were designed specifically to monitor global stratospheric aerosols. Examples of the global aerosol climatology that is evolving for the first time will be described, along with a brief description of some applications in stratospheric aerosol studies.

2.4.1 Present Spacecraft Experiments

The first unmanned satellite-borne instrument for measuring stratospheric aerosols is the Stratospheric Aerosol Measurement II (SAM II) which was launched on the Nimbus 7 spacecraft, 24 October 1978. SAM II is designed to measure stratospheric aerosol extinction properties in the 1.0 μm wavelength region. The 955-km nearly circular orbit for Nimbus 7 is a high-noon sunsynchronous one, that is, when the subsatellite point crosses the equator it is local noon or midnight. This orbital geometry restricts SAM II measurements to latitude bands of approximately $64°$-$80°$ in both the northern and southern hemisphere covering just about the entire arctic and antarctic regions. The second instrument placed into earth orbit is the Stratospheric Aerosol and Gas Experiment (SAGE) which was launched on a dedicated Application Explorer Mission-B (AEM-B) spacecraft, 18 February 1979. The 600-km nearly circular orbit for AEM-B has an orbit plane that is highly precessing. It provides the SAGE instrument, therefore, with measurements distributed around the earth between the latitudes of approximately $79°$N and $79°$S. The SAGE instrument measures stratospheric aerosol extinction in the 1.0 and 0.45 μm wavelength regions, as well as absorption due to ozone in the Chappuis band (0.6 μm) and nitrogen dioxide in the 0.385 to 0.45 μm region. Figure 2.17 illustrates the geometry associated with the solar extinction technique, showing how the satellite measures solar intensity through different atmospheric layers at different altitudes during a single sunrise or sunset event encountered by the satellite. As the satellite approaches the shadow side of Earth (spacecraft sunset), light from the sun to the satellite passes through successively lower altitude air masses until the sun is occulted by clouds or the Earth's surface. Solar intensity measurements during this transient period provide a complete altitude scan of the atmosphere. During a satellite sunrise event, the measurement sequence is reversed. Typical measurements will cover altitudes from the ground or cloud-top to above 250-km altitude where there is no atmospheric attenuation. In fact there is no appreciable extinction above about 60 km over the SAM II and SAGE spectral regions. This high-altitude solar intensity data provide a convenient self-calibration for occultation instruments and is one reason why this technique is so successful. As seen by the spacecraft, the sun rises or sets at a vertical rate of about 3 km s^{-1}. A scan of the stratosphere, therefore, takes about 20 s and the entire atmosphere (300 km) is scanned in only 100 s; thus, occultation instruments need to maintain their calibration or through-

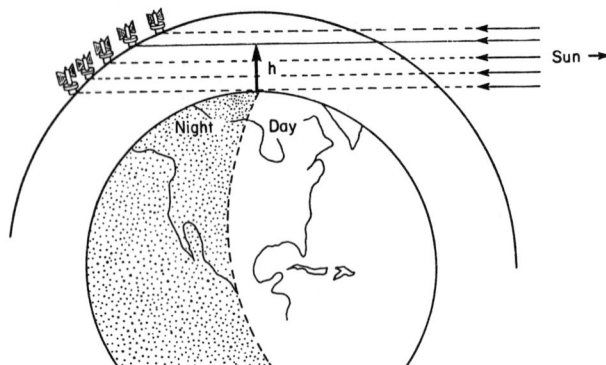

Fig.2.17. Geometry for solar extinction measurements with satellite-mounted radiometers. Different layers of the atmosphere at tangent height h are successively sampled during the occultation event [2.139]

put and gain characteristics for only a short time. This characteristic coupled with a zero-atmosphere calibration for each event and an inherent high signal-to-noise ratio for a solar source make this a powerful remote sensing technique.

The equation describing the relationship between aerosol extinction properties and the satellite's measurements is

$$H_\lambda(t) = \int_{\Delta\lambda} \int_{\Delta\Omega} F_\lambda(\theta,\phi) S_\lambda(\theta,\phi,t) T_\lambda(\theta) d\Omega d\lambda \qquad (2.5)$$

where $H_\lambda(t)$ is the instantaneous irradiance measured by the instrument at center wavelength λ and time t, $F_\lambda(\theta,\phi)$ is the radiometer's field of view function, ϕ is the azimuthal angle, Ω is the solid angle, $T_\lambda(\theta)$ is the transmission of the atmosphere at λ as a function of view angle θ, and $S_\lambda(\theta,\phi)$ is the extraterrestrial solar radiance profile. Effects due to atmospheric refraction have to be included in computing $S_\lambda(\theta,\phi)$. The transmission function $T_\lambda(\theta)$, with the change of variable from θ to tangent height h_t, is given by the Bouguer law as

$$T_\lambda(h_t) = \text{Exp} -\left[\int \beta_\lambda(h) d\rho_\lambda(h)\right] \qquad (2.6)$$

where $\beta_\lambda(h)$ is the total extinction coefficient of the atmosphere versus altitude h for wavelength λ, and $\rho_\lambda(h)$ is the sun ray optical pathlength. In general, the total extinction coefficient β_λ at each altitude is composed of contributions from aerosol and molecular scattering and specific molecular absorptions. At the four SAGE wavelengths

$$\beta_\lambda = \beta_{nd}(\lambda) + \beta_{NO_2}(\lambda) + \beta_a(\lambda) + \beta_{O_3}(\lambda) \qquad (2.7)$$

where $\beta_{nd}(\lambda)$ is the Rayleigh extinction coefficient at λ; $\beta_{O_3}(\lambda)$, $\beta_{NO_2}(\lambda)$, and $\beta_a(\lambda)$ are, respectively, ozone, nitrogen dioxide, and aerosol extinction coefficients at λ. At the 1.0 μm wavelength common to both SAM II and SAGE, β_λ is composed solely of Rayleigh and aerosol contributions. Thus aerosol extinction properties at 1.0 μm can be retrieved independently from satellite measurements at

1.0-μm wavelength together with appropriate meteorological data. Details of the technique used are given in CHU and McCORMICK [2.143].

a) *SAM II*

The SAM II instrument is a single-spectral channel sun photometer with passband centered at 1.0-μm wavelength with a bandwidth at half maximum of about 0.02 μm. Solar radiation is reflected by a scan mirror and collected by a Cassegrainian telescope to produce an image of the solar disk on the telescope's focal plane. On the focal plane is a circular aperture that defines a 0.6 arcmin instantaneous field of view (IFOV). This IFOV provides an atmospheric vertical resolution on the horizon of approximately 0.5-km altitude. Sunlight passing through the aperture is directed by a lens through a bandpass filter to a silicon photodiode (science detector).

Immediately before a satellite sunrise or sunset event, the SAM II instrument is activated by a sun presence sensor indicating that the sun is within the instrument's field of view. The instrument then locks onto the sun in azimuth and scans in elevation until the sun is acquired by the IFOV. The scan mirror then scans vertically, with respect to the Earth's horizon, across the solar disk at a rate of 15 arcmin per second, reversing the scan direction each time a sun edge crossing occurs. The solar edge crossings are detected by edge detectors placed on the focal plane on either side of the circular aperture that defines the IFOV. The output from the science detector is sampled at 50 HZ and digitized with a 10-bit A to D converter. The digital data are stored on-board the satellite together with auxiliary data about the instrument, and are later transmitted to ground stations for analysis.

b) *SAGE*

The SAGE instrument is an advanced version of SAM II which can simultaneously measure solar intensity at four separated spectral wavelength bands. Its configuration is basically identical to the SAM II instrument except that it has a spectrometer which utilizes a holographic grating for spectral discrimination. Sunlight is dispersed into four spectral channels, with wavelengths centered at 1.0, 0.6, 0.45, and 0.385 μm. Four separated pinphotodiode detectors are positioned on the Rowland circle to detect first-order dispersed light. The additional channels at shorter wavelengths on SAGE provide stratospheric ozone and nitrogen dioxide concentration profiles, and aerosol extinction profile properties at 0.45 μm. The operational sequence of the SAGE instrument is identical to that of SAM II's except the solar irradiance is sampled at a rate of 64 HZ and digitized to 12-bit accuracy in order to accommodate the larger signal dynamic range requirement for the shorter wavelength channels.

c) Ground-Truth Programs

The ground-truth programs for the SAM II and SAGE experiments utilize a coordinated set of experiments, in which aerosol content and other properties of a given air mass were measured with a variety of different instruments nearly simultaneously with the satellite overpass. Of all the stratospheric aerosol experiments available, only lidar and balloon-borne dustsonde systems can provide the necessary high-resolution vertical profiles of aerosol properties with reasonable accuracy. The dustsonde measures concentrations of aerosols in two different size ranges from the ground to about 28-km altitude. Both the lidar measurements and the dustsonde measurements have, over the years, demonstrated their reliability and comparability in measuring stratospheric aerosols (see Fig.2.8 and [2.104,110,119,128,129]). The SAM II and SAGE ground-truth programs thus incorporated the lidar and dustsonde systems as their prime ground-truth sensors for validation of the satellite results. An airborne ruby and Neodymium-YAG lidar system was developed at NASA Langley Research Center and has performed ground-truth measurements at locations not easily accessible by other means, or during inclement weather conditions. In addition to lidar and the University of Wyoming dustsonde sensors, other measurements such as an in situ quartz crystal microbalance impactor and a polar nephelometer, have been included in the aerosol part of the SAM II and SAGE ground-truth programs. These complemented the lidar and dustsonde measurements with some additional aerosol characteristics at a few altitudes. Details of these ground-truth plans are given in [2.144,145]. A number of sensors have also been flown in NASA's Aerosol Climatic Effects Program [2.146] aboard a NASA U-2 aircraft in support of the SAM II and SAGE validation [2.130].

Major ground-truth experiments that have been carried out through 1980 for SAM II and SAGE aerosol measurements are listed in [2.144,145]. There are also numerous other associated experiments that have been carried out in Japan, England, Germany, Belgium, France, and Australia, by participating international ground-truth experiment teams [2.147]. The first published results of the successful validation of the SAM II data are given in [2.128].

2.4.2 Orbital Considerations

Measurement opportunities for satellite solar extinction techniques are limited by the satellite orbit and Earth-sun geometry [2.148]. For typical satellite orbital periods of approximately 100 minutes, there are 30 measurement events (sunrises and sunsets) per day (about 11,000 per year).

The latitudinal coverages of the SAM II and SAGE measurements are illustrated in Fig.2.18 for the first year after launch. Each day 15 sunrises and 15 sunsets occur at these latitudes, separated by about 24 degrees in longitude. The Nimbus 7 satellite is in a nearly circular, retrograde orbit with inclination of $99°$. The latitudinal coverage is repeated four times a year with minimum latitudes measured

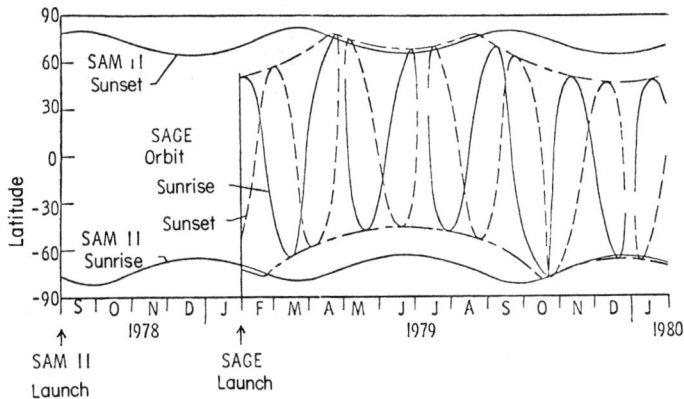

Fig.2.18. Latitudinal coverages of SAM II and SAGE measurements for the one year period after launch. The coverage for SAM II is limited to the northern and southern polar regions, while SAGE's coverage extends from 79°S to 79°N. Both instruments make about 11,000 measurements per year. [2.139]

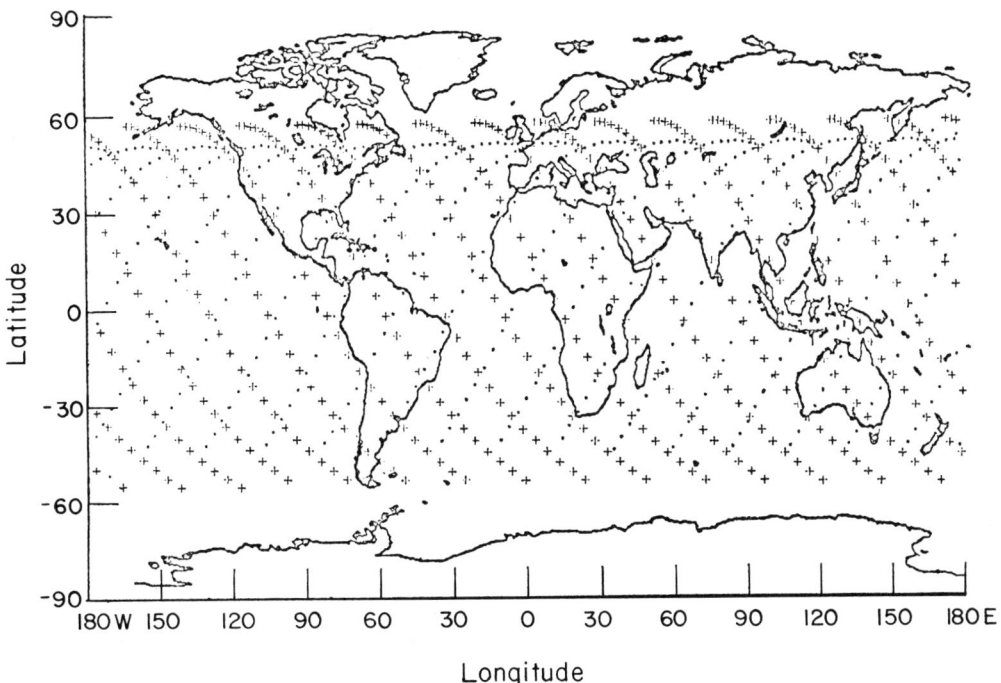

Fig.2.19. SAGE measurement locations for February 1979. Successive events are shifted slightly in longitude and latitude. The + symbols represent sunset events, and the · symbols represent sunrise events. [2.139]

at the solstices and maximum latitudes at the equinoxes. The limited latitudinal coverage for Nimbus 7 satellite is characteristic of sun-synchronous orbits and can be varied somewhat by changing the launch and, therefore, the equatorial crossing time (see [2.148] for details). The AEM-B satellite is in a non-sun-synchronous orbit with an inclination of 55°. This highly precessing orbit can repeatedly cover 120 degrees of latitude every two to three weeks with a total coverage over the year from 79°S to 79°N. Figure 2.19 gives an idea of the geographic coverage of SAGE for one month. The tangent points of the sunrise and sunset events during February 1979 are shown. For non-sun-synchronous orbits, like the one SAGE is in, the total latitudinal coverage is inversely proportional to the frequency of repeating latitude. Thus it is important to consider the trade-off between total latitudinal coverage and the frequency of repeating latitude cycles or, equivalently the number of profiles obtained in any given latitude band in a given amount of time. The combined coverages of the SAM II and SAGE instruments, as illustrated in Fig.2.18, provide some measurement overlap which has been used for comparisons between SAM II and SAGE results. The overall coverage from both SAM II and SAGE is nearly global, covering approximately 98% of the global surface. Details of tailoring orbits for occultation measurements as was done for SAGE are given in HARRISON et al. [2.148].

2.4.3 Results Obtained with SAM II

The basic SAM II data product is the vertical aerosol extinction profile obtained during each measurement opportunity. There are a total of approximately eleven thousand profiles obtained for each year that are evenly divided between the north and south polar regions. The SAM II aerosol extinction profiles can be analyzed to determine latitudinal, longitudinal, and temporal variations in the stratospheric aerosol layer. Seasonal and hemispheric differences in the aerosol can also be evaluated.

Figure 2.20 [2.149] illustrates average weekly profiles of aerosol extinction at 1.0 μm wavelength obtained by SAM II from 10 December 1978 to 20 January 1979, corresponding to a latitude coverage for this period from 65.2°N to 67.5°N. The horizontal bars associated with each profile denote one standard deviation in the weekly data set, and the arrow denotes the average tropopause height for the week at those latitudes. Accompanying each weekly aerosol profile, is the average weekly temperature profile with its one standard deviation also, which is an indication of the data fluctuation. The temperature data are supplied by the U.S. National Weather Service and correspond to the location and time of the SAM II profiles. Tropospheric clouds are easily recognized by the large fluctuations near and below the tropopause. Clouds above the average tropopause are due to vertical variations with longitude of the tropopause. The mean stratospheric aerosol extinction coefficient at 1 μm is seen to be about 1 to 2×10^{-4} km^{-1} between tropopause and 20 km. Figure 2.21

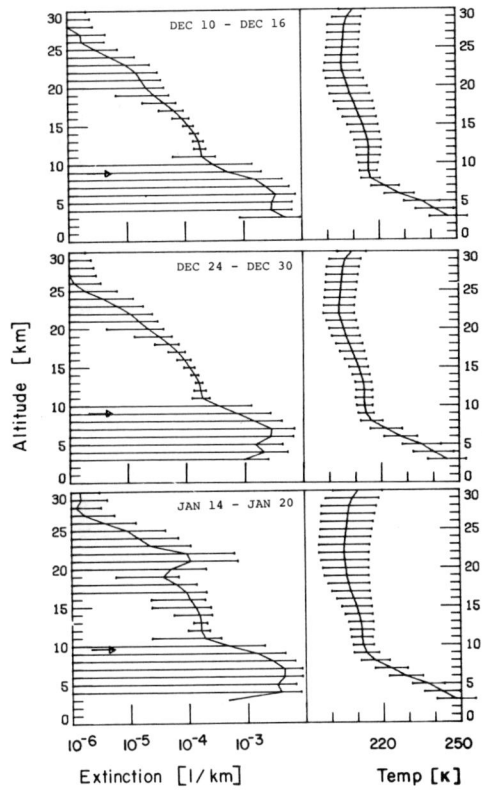

Fig.2.20. Average weekly aerosol extinction profiles at 1.0 μm as measured by SAM II from 10 December 1978 to 20 January 1979, and the corresponding average weekly temperature profiles and the horizontal arrows the tropopause altitude. The horizontal bars denote one standard deviation. The latitudinal coverage during this time period is from 64.9°N to 67.9°N. [2.149]

[2.149] shows altitude-versus-longitude isopleths of aerosol extinction for 19 December 1978 at a latitude of 69.4°N. The extinction isopleth values are scaled by a factor of 10^{-5} for aerosol extinction in units of km^{-1}. The tic marks on the abscissas indicate the longitude of the individual SAM II measurements used to construct the isopleth. Tropospheric clouds with high extinction values near an altitude of 10 km, and the presence of longitudinal variations of the stratospheric aerosol layers up to 20-30 km altitude, are easily seen. The corresponding temperature isopleths are shown in the lower graph. The vertical line near the center of each graph marks the zero meridian. Figures 2.20,21 illustrate typical background aerosol values during this period with peak values occurring from about 12 to 17 km with a nearly constant extinction of about 10^{-4} km^{-1}.

The time histories for one year of stratospheric aerosols for the arctic and antarctic regions are illustrated in Figs.2.22,23 [2.150]. Figure 2.22b shows isopleths of weekly averaged aerosol extinction, using the same units as in Fig.2.21, as a function of altitude and time, for the northern hemisphere from 29 October 1978 through 27 October 1979. The dashed quasi-horizontal line near 10 km shows the position of the average tropopause for each week. Figure 2.22c shows the corresponding isopleths of temperature in degrees Kelvin. Figure 2.23 displays the equivalent

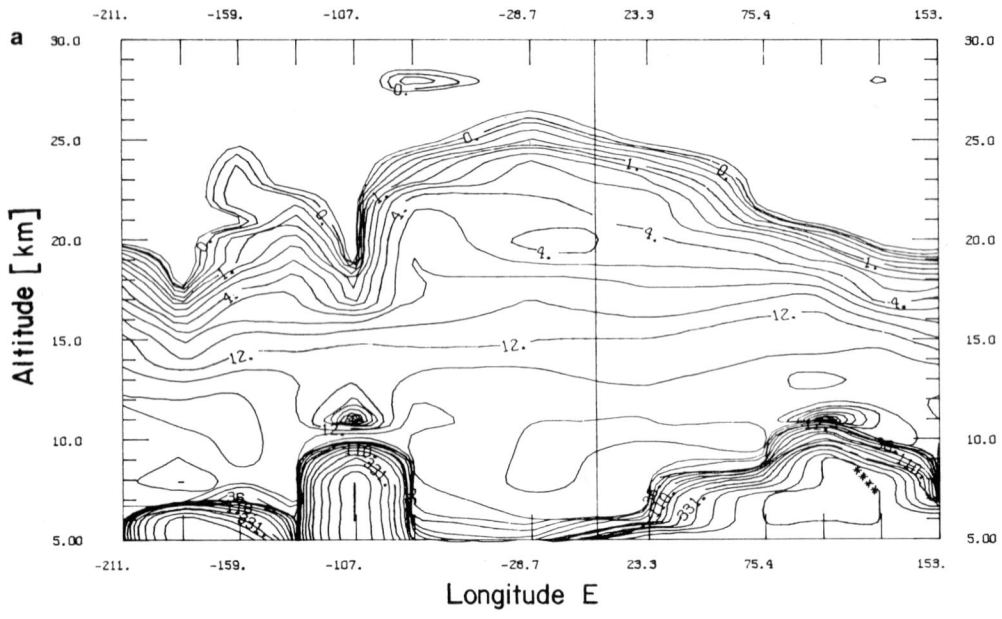

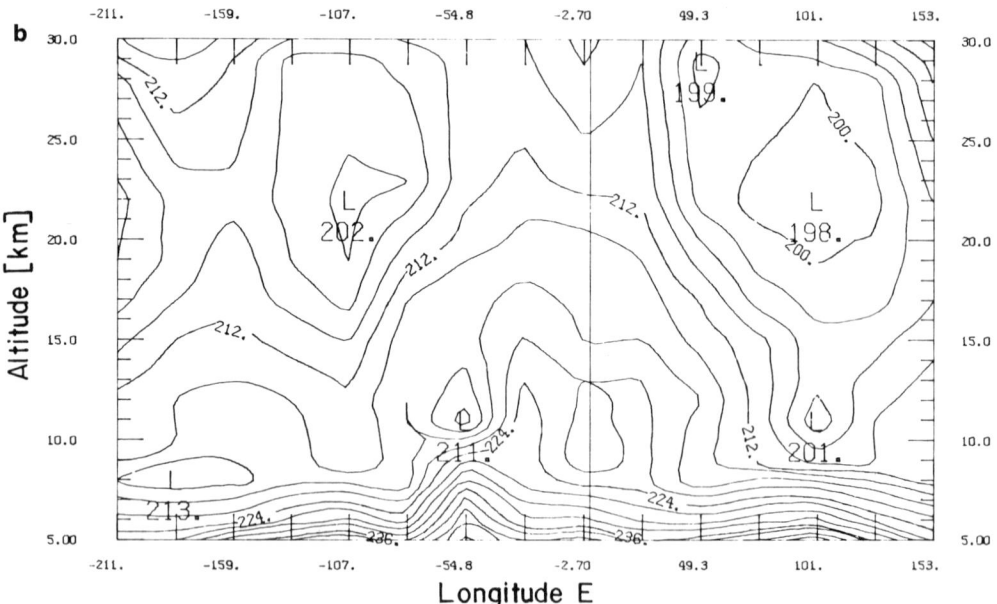

Fig.2.21. (a) Isopleths of aerosol extinction at 1.0 μm versus altitude and longitude for 19 December 1978 at latitude 64.9°N. Isopleths are scaled by 10^{-5} km^{-1}. The vertical line indicates the zero meridian. (b) Isopleths of temperature in Kelvins, at 3° intervals, for the same time period and locations as in (a). [2.149]

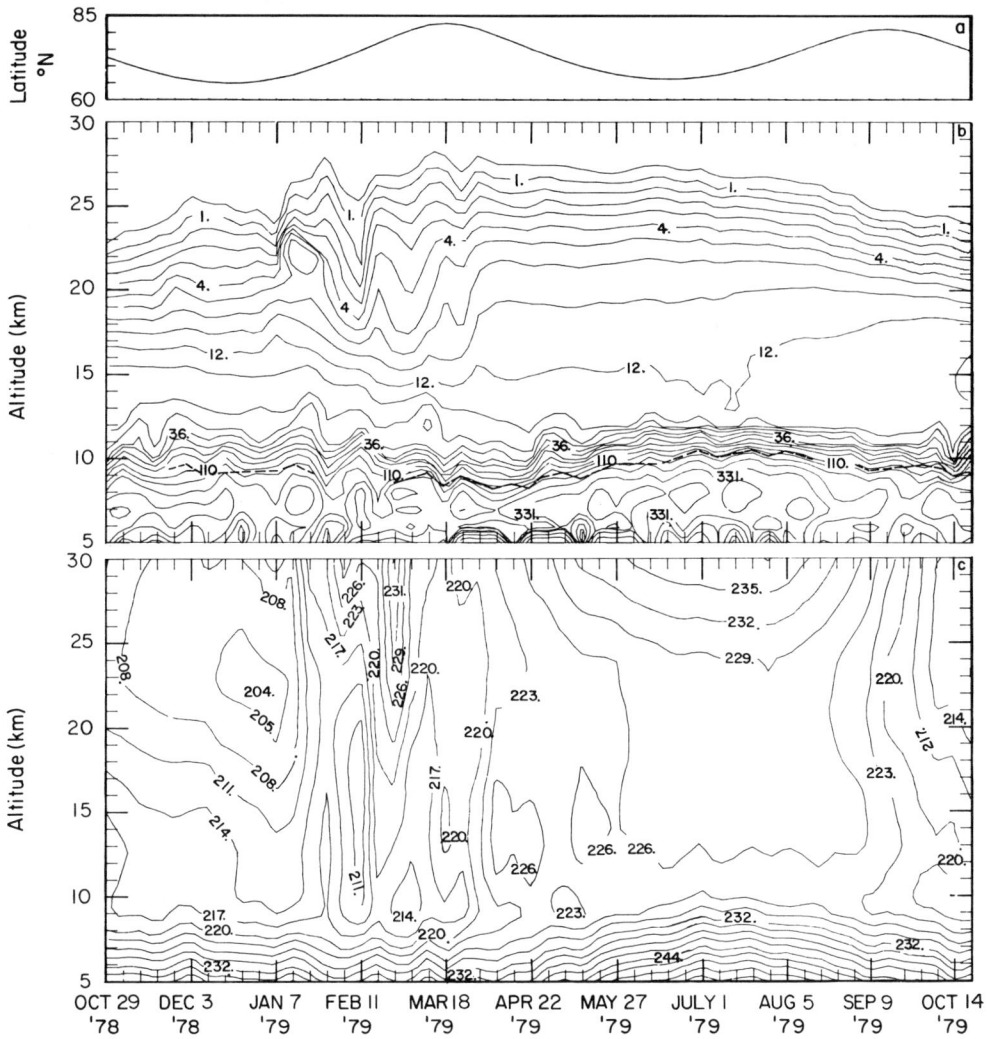

Fig.2.22a-c. SAM II measurements in the northern hemisphere. Data show isopleths of weekly averaged aerosol extinction at 1.0 μm, where the date marked on the horizontal axis is the first day of the week to which the average value corresponds. (a) Latitude of SAM II measurements; (b) Aerosol extinction isopleths in units of 10^{-5} km^{-1}. Dashed line shows averaged tropopause heights; (c) Corresponding temperature field at the location of aerosol measurements. [2.150]

information for the southern hemisphere, but displaced by 26 weeks in order to facilitate a comparison of the seasonal behavior between the two hemispheres.

The SAM II results have so far indicated that the seasonal behavior of the stratospheric aerosol in the two polar regions are similar and that there are strong correlations with temperature. In the winter period, increased aerosol extinctions are found in both hemispheres. The large increases in extinction occurring during

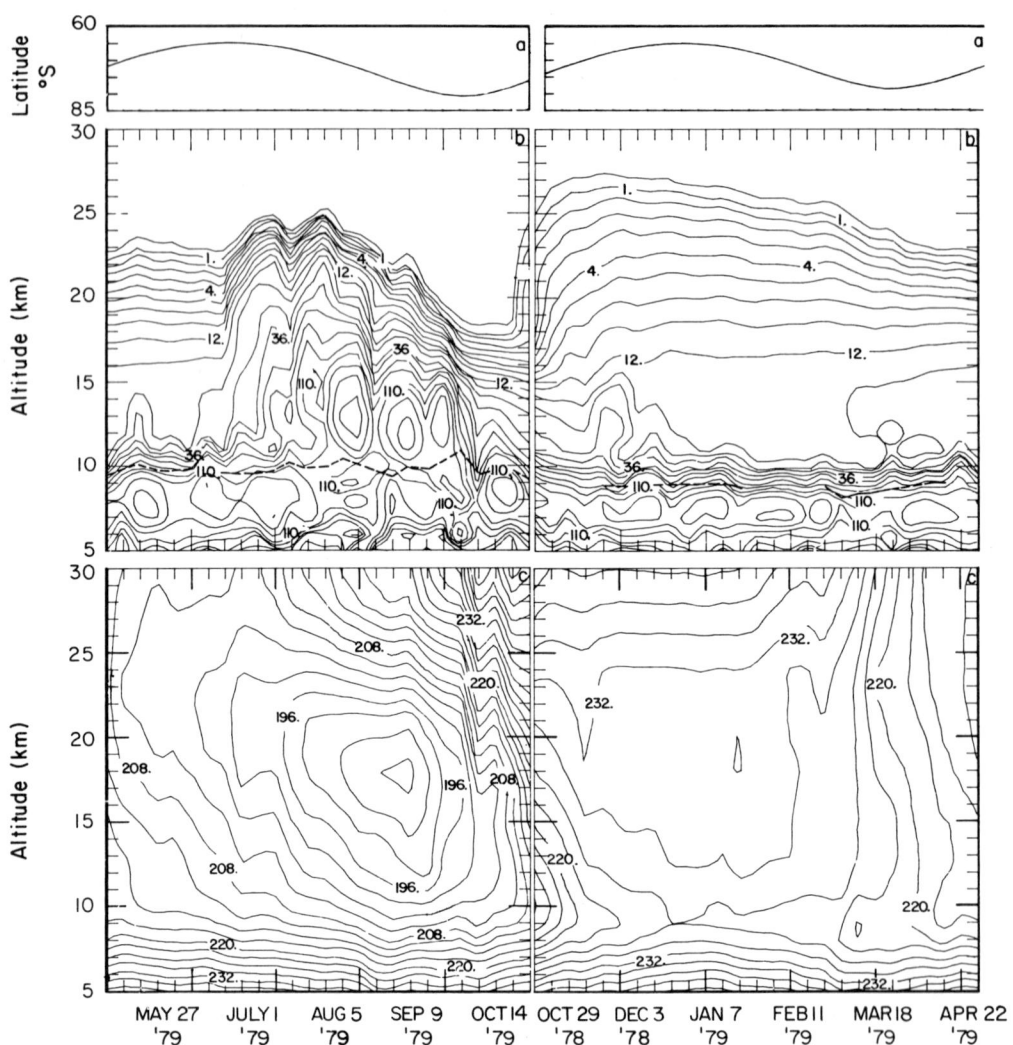

Fig.2.23a-c. SAM II measurements in the southern hemisphere. Data show isopleths of weekly averaged aerosol extinction at 1.0 μm, where the date marked on the horizontal axis is the first day of the week to which the average value corresponds. (a) Latitude of SAM II measurements; (b) Aerosol extinction isopleths in units of 10^{-5} km^{-1}. Dashed line shows averaged tropopause height; (c) Corresponding temperature field at the location of the aerosol measurements. Fig.2.22 covers the same time interval as Fig.2.23 but the latter was divided into two halves which were interchanged so that similar seasons in the two figures are aligned. [2.150]

particularly cold temperatures are manifestations of stratospheric clouds, thought to be ice [2.151]. These are occasionally sighted in the arctic winter but are ubiquitous in the antarctic winter [2.152]. Toward the end of winter, the aerosol layer descends in both polar regions, followed by a rapid ascent to a height of about 25 km in early spring. Following this period the top of the aerosol layer falls steadily throughout the summer, and stays nearly constant through the fall

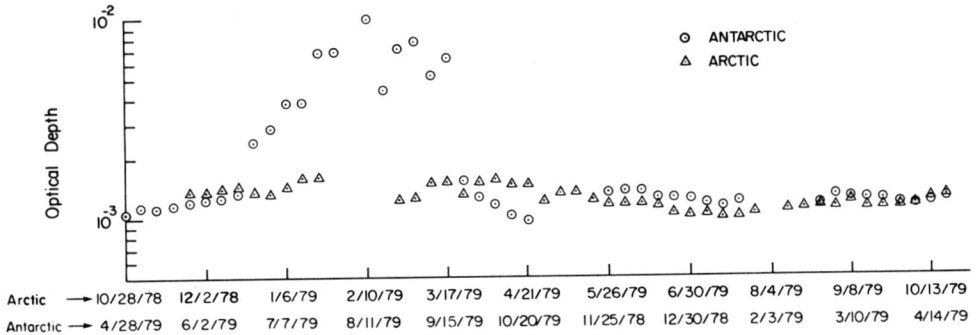

Fig.2.24. Weekly averaged optical depths at 1.0 μm as measured by SAM II. Note that antarctic values are displaced by six months as in Fig.2.23. [2.150]

season. Figure 2.24 shows the corresponding one-year optical depth data, where the values for the southern hemisphere are again displaced by six months. These optical depths were obtained by integrating SAM II extinction data from 2 km above the tropopause heights. The stratospheric aerosol optical depth at 1 μm in the two hemispheres is similar, about 10^{-3}, except during the winter period where the stratospheric clouds contribute. Note that the antarctic optical depth is enhanced by as much as an order of magnitude due to these clouds.

SAM II observed stratospheric clouds in the northern hemisphere during the winter of 1978-1979, primarily in the month of January. These observations showed enhancements in the 16-22 km altitude range to be one to two orders of magnitude larger than the normal background values. The temperature profiles for each observation show a very cold stratosphere with temperature as low as 192 K. A total of twelve observations of the stratospheric clouds were made in the northern hemisphere during this winter season. For the southern hemisphere where the winter period begins in early July, occurrences of stratospheric clouds were found to be much more prevalent. In the arctic they are highly localized; that is, they are observed on one or at most three subsequent orbits; in the antarctic, however, they are usually observed during several subsequent orbits and often persist for several days to several weeks. A total of 538 stratospheric cloud observations in the antarctic were counted during a period of 14 weeks from 24 June through 29 September 1979. The prevalence of stratospheric clouds in the antarctic is illustrated in Fig.2.23b where high extinction values occur persistently (compared with the arctic winter) over the altitude range from 10 to 20 km during the entire winter period. During the same period, stratospheric temperatures were consistently below 200 K. Based on thermodynamic considerations, the formation process of stratospheric clouds as sighted by the SAM II instrument has been analyzed by HAMILL et al. [2.151]. They have shown that a reasonable formation process is the growth and dilution of the background stratospheric aerosol particles due to the absorption of water vapor, followed by freezing and the subsequent growth of an ice shell around the frozen sulfate core.

2.4.4 Results Obtained with SAGE

Like SAM II, the basic SAGE data product consists of 1.0 μm aerosol extinction profiles but, in addition, SAGE data include aerosol extinction at 0.45 μm and ozone and nitrogen dioxide concentration or mixing ratio vertical profiles. The SAGE aerosol data (79°S to 79°N) can be used to analyze temporal variations in the stratospheric aerosol layer on a near global scale. Moreover, transient perturbations such as the injection of volcanic material and its subsequent dispersion have been determined from the SAGE aerosol data.

Figure 2.25 plots an isopleth of the zonal mean aerosol extinction at 1.0 μm wavelength as a function of altitude and latitude collected by SAGE in April 1979 between the latitudes 55°S and 55°N. About 900 individual aerosol vertical profiles were used to construct this plot. The dashed line between 10 and 17 km denotes the mean tropopause heights for these latitudes during this time period. The high extinction values immediately below the tropopause levels represent occurrences of tropospheric clouds. The stratospheric aerosol layer is again shown to have extinction contour lines of approximately 10^{-4} km^{-1} and generally follow the trend of the tropopause level. This result confirms earlier findings with balloon soundings and lidar observations, showing a vertical shift of the stratospheric aerosol layer at different latitudes that is correlated strongly with tropopause height [2.80,118]. When converted to extinction ratio (dividing by molecular extinction), these data are consistent with a tropical source for stratospheric aerosols. This April 1979 aerosol data set represents near-background stratospheric aerosol conditions since the last major volcanic stratospheric enhancement occurred in 1974 from Volcán de Fuego [2.120]. An extinction value of 10^{-4} km^{-1} at 1.0 μm wavelength, therefore, represents typical background stratospheric aerosol conditions.

The ability to observe volcanic injections of material into the stratosphere by SAGE was first demonstrated during April 1979, when the Soufrière volcano on St. Vincent (13.3°N, 61.2°W) erupted several times, sending considerable material into the stratosphere. SAGE observations shortly after the volcanic eruption indicated enhanced aerosol extinction at about an altitude of 20 km at locations near the volcano and extending northeast over the Atlantic Ocean and the western shore of Africa [2.153]. Figure 2.26a shows a 1.0 μm background aerosol extinction profile on 24 April 1979. Figure 2.26b taken on the same day, shows an enhancement in the normal profile due to the volcanic plume. The profile shows an extinction maximum at 20.4 km, which is approximately four times greater than the normal value. Figure 2.26c shows the SAGE observations during the period from 19 to 30 April 1979, identifying volcanic plumes at eight separate locations, spanning latitudes from 14° to over 35°N. The horizontal dashed lines represent the latitudes covered by SAGE for each day as indicated. Trajectories analyzed for the Soufrière volcanic plumes, using meteorological data, show good agreement with the SAGE observations. McCORMICK et al. [2.153] used these SAGE observations (with conversion of extinction

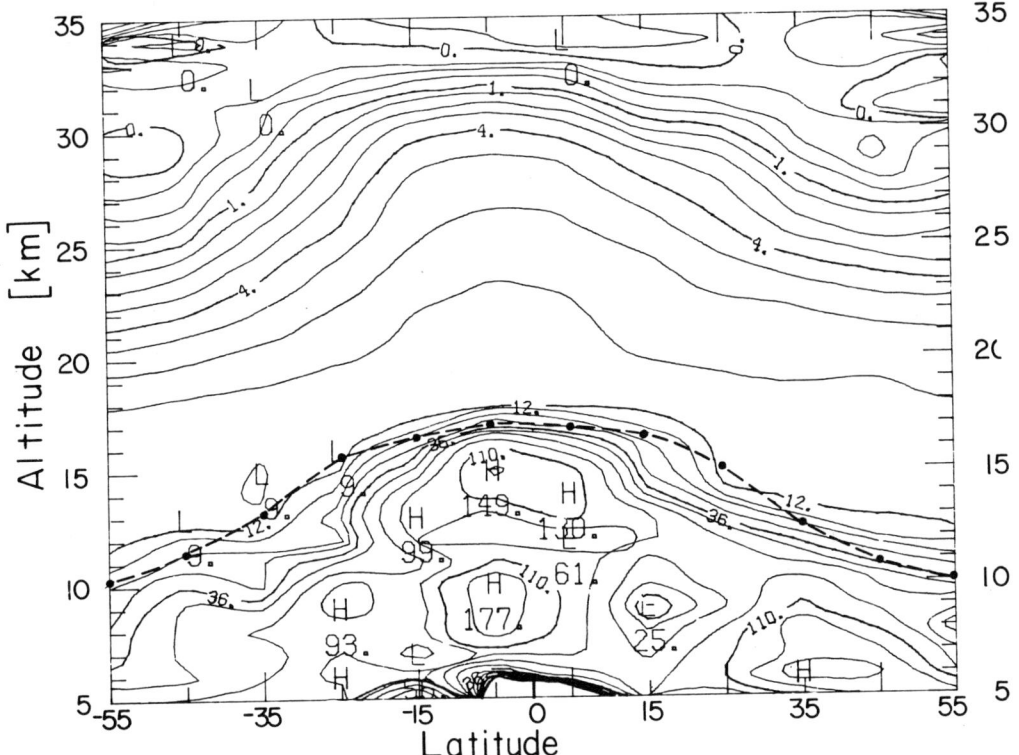

Fig.2.25. Isopleths of aerosol extinction at 1.0 μm versus altitude and latitude for the month of April 1979 as measured by SAGE. Isopleths are scaled by 10^{-5} km^{-1}. The broken line indicates the average tropopause height

profiles to mass via an optical model), and estimated Soufrière's contribution to the stratospheric aerosol mass loading to be approximately 0.5% of the total global value.

In addition to the observations of Soufrière, SAGE has observed at least two other stratospheric volcanic injection events; the Sierra Negra volcano (0.8°N, 91°W) which erupted in late November 1979, and the Mt. St. Helens volcano (46°N, 125°W) which erupted violently on 18 May 1980. Both eruptions were accompanied by large amounts of volcanic materials (ash and gas) injected into the stratosphere. Since Mt. St. Helens is located on the continental U.S., intensive observations of the eruption events have been conducted, making this probably the best-studied volcanic eruption in history [2.154,155]. SAGE observations on the enhancement of stratospheric aerosols due to the 18 May 1980 Mt. St. Helens' eruption are illustrated in Fig.2.27. Latitudinal variations of the zonal mean of stratospheric aerosol extinction at 1.0 μm, during the period 1 July to 12 August 1980, are shown as a function of altitude. The majority of the volcanic material is seen to have moved northward to higher latitudes, especially directly above the tropopause. Using the same optical conversion as in [2.153], McCORMICK [2.156] estimated that

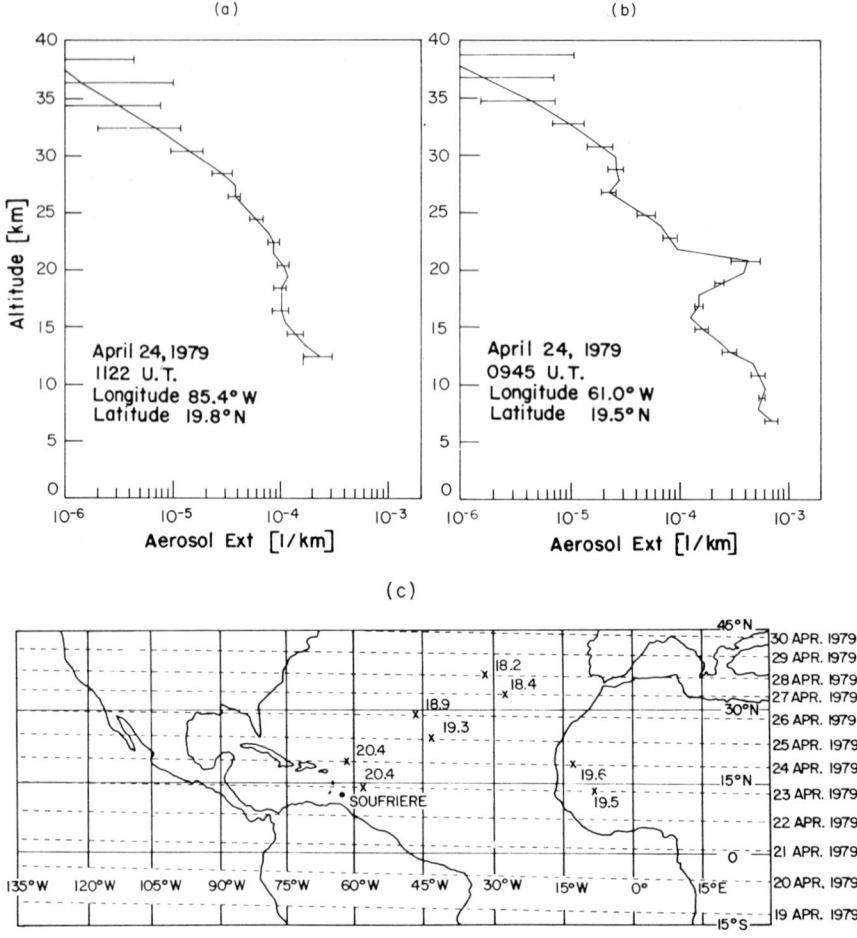

Fig.2.26. (a) Background aerosol extinction profile at 1.0 μm as measured by SAGE; (b) volcanically enhanced aerosol profile; (c) map showing SAGE measurement coverage during the Soufrière April 1979 eruptions. SAGE events showing enhanced extinction are marked by crosses with the layer heights in km shown alongside each cross. [2.153]

the 18 May 1980 eruption increased the northern hemisphere background stratospheric aerosol by a factor of two, putting about 0.32×10^6 tons of new material into the stratosphere.

2.4.5 Ground-Truth Comparisons

As stated in Sect.2.4.1, SAM II and SAGE ground-truth programs rely on simultaneously measured data from ground-truth sensors, such as airborne lidar and balloon-borne dustsonde, to validate the satellite measurements. The satellite data consist primarily of aerosol extinction, while the ground-truth sensors measure different microproperties, such as backscattering (lidar) or particle density (dustsondes). In order to accomplish the validation process, therefore, it is necessary to convert

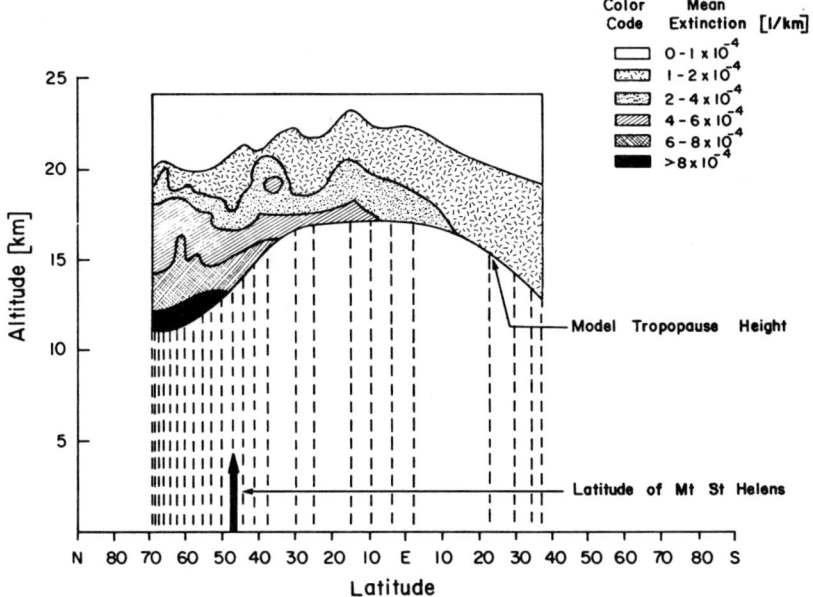

Fig.2.27. Altitude versus latitude map of aerosol extinction at 1.0 μm as measured by SAGE from 1 July to 12 August 1980, showing enhancements caused by the eruptions of Mt. St. Helens. The vertical broken lines indicate latitude locations of the SAGE measurements. Normal background would be $1-2 \times 10^{-4}$ km^{-1}

the ground-truth measured data into aerosol extinction coefficients through the use of a realistic optical model for atmospheric aerosols. In addition, the uncertainties associated with the conversion process must be understood and included. A model of background stratospheric aerosols, including their optical properties and variability, has been developed under the SAM II ground-truth program [2.129]. The model is constructed from empirical data and consists of three regions, the upper troposphere, the tropopause region, and the inner stratosphere.

Results from the Sondrestrom, Greenland, ground-truth experiments for SAM II are described in Sect.2.3.3 and are shown in Fig.2.14. The three profiles for both days agree to within their respective uncertainties at all heights above the tropopause. Near the tropopause, the satellite measures significantly higher extinction compared to the value inferred from dustsonde. This is caused by the occurrence of patchy cirrus clouds along the satellite's long horizontal view path. Similar results were obtained during ground-truth experiments conducted at Poker Flat, Alaska, 16 to 19 July 1979. Coincident measurements from SAM II and SAGE were obtained during this period when the coverages of SAM II and SAGE overlapped. Good agreement was obtained between SAM II and SAGE data, as well as between the satellite data and the data from the ground-truth sensors. This is typical of all the other ground-truth experiments associated with SAM II or SAGE measurements and provides strong support for the validity of the SAM II and SAGE extinction data and their uncertainty estimates, which were derived from an independent error analysis.

2.4.6 Applications

The global stratospheric aerosol data being provided by SAM II and SAGE provide a highly detailed picture of the stratospheric aerosol distribution. This continuous and global set can be used in a number of atmospheric studies such as the gas-to-particle conversion of sulfur gases, the global dispersion of effluents from a volcanic injection, radiative effects of polar stratospheric ice clouds, the global mass of the stratospheric aerosol, the use of aerosol as tracers for atmospheric motions, and the potential climate effects of aerosols as discussed in Chap.5.

References

2.1 E. Sagawa, T. Itoh: Geophys. Res. Lett. *4*, 29-32 (1977)
2.2 W. Jaeschke, R. Schmitt, H.-W. Georgii: Geophys. Res. Lett. *3*, 517-519 (1976)
2.3 H.-W. Georgii, F.X. Meixner: J. Geophys. Res. *85*, 7433-7438 (1980)
2.4 J. Stauff, W. Jaeschke: Atmos. Environ. *9*, 1038-1039 (1975)
2.5 J.F. Vedder, B.J. Tyson, R.B. Brewer, C.A. Boitnott, E.C.Y. Inn: Geophys. Res. Lett. *5*, 33-36 (1978)
2.6 E.C.Y. Inn, J.F. Vedder, B.J. Tyson, D. O'Hara: Geophys. Res. Lett. *6*, 191-193 (1979)
2.7 A.L. Schmeltekopf, P.D. Goldan, W.J. Harrop, T.L. Thompson, D.L. Albritton, M. MacFarland, A.E. Sapp, W.R. Henderson: Rev. Sci. Instrum. *47*, 1479-1485 (1976)
2.8 J.F. Vedder, E.C.Y. Inn, B.J. Tyson, D. O'Hara: J. Geophys. Res. *86*, 7363-7368 (1981)
2.9 E.C.Y. Inn, J.F. Vedder, D. O'Hara: Geophys. Rev. Lett. *8*, 5-9 (1981)
2.10 W.G. Mankin, M.T. Coffey, D.W.T. Griffith, S.R. Drayson: Geophys. Res. Lett. *6*, 853-856 (1979)
2.11 E.C.Y. Inn, J.F. Vedder, E.P. Condon, D. O'Hara: Science *211*, 821-823 (1981)
2.12 P.D. Goldan, W.C. Kuster, D.L. Albritton, A.L. Schmeltekopf: J. Geophys. Res. *85*, 413-423 (1980)
2.13 R.P. Turco, P. Hamill, O.B. Toon, R.C. Whitten, C.S. Kiang: J. Atmos. Sci. *36*, 699-717 (1979)
2.14 P.J. Maroulis, A.L. Torres, A.R. Bandy: Geophys. Res. Lett. *4*, 510-512 (1977)
2.15 F.J. Sandalls, S.A. Penkett: Atmos. Environ. *11*, 197-199 (1977)
2.16 N.D. Sze, M.K.W. Ko: Nature London *280*, 308-310 (1979)
2.17 O.B. Toon, R.P. Turco, R. Whitten, P. Hamill: Geophys. Res. Lett. *8*, 23-25 (1981)
2.18 P. Maroulis, A.R. Bandy: Geophys. Res. Lett. *1*, 681-684 (1980)
2.19 C.E. Junge, C.W. Chagnon, J.E. Manson: J. Meteorol. *18*, 81-108 (1961)
2.20 R.D. Cadle, C.S. Kiang: Rev. Geophys. Space Phys. *15*, 195-202 (1977)
2.21 E.K. Bigg, A. Ono, W.J. Thompson: Tellus *22*, 550-563 (1970)
2.22 D.C. Woods, R.L. Chuan: EOS *60*, 834 (1979)
2.23 E.K. Bigg, Z. Kviz, W.J. Thompson: Tellus *23*, 247-259 (1971)
2.24 R.L. Chuan: In *Fine Particles, Aerosol Generation, Measurement and Sampling Analysis*, ed. by B.Y.H. Liu (Academic, San Francisco 1976) pp.763-775
2.25 P.W. Hodge: "Sampling Dust from the Stratosphere," Smithson. Contrib. Astrophys. *5*, (1961)

2.26 J.P. Friend: Tellus *18*, 465-473 (1966)
2.27 G.V. Ferry, H.Y. Lem: In *Proc. 2nd Int. Conf. on Environ. Impact of Aerospace Operations in the High Atmos.*, ed. by R. Collis (Amer. Meteorol. Soc., Boston, MA 1974) pp.27-33
2.28 D.E. Brownlee, P.W. Hodge, W. Bucher: NASA Spec. Publ. SP-*319*, 291 (1973)
2.29 N.H. Farlow, D.M. Hayes, H.Y. Lem: J. Geophys. Res. *82*, 4921-4929 (1977)
2.30 H.Y. Lem, N.H. Farlow: "Efficiency of Aerosol Collection on Wires Exposed in the Stratosphere"; Tech. Memo. NASA 81147, Ames Research Center (1979)
2.31 N.H. Farlow, G.V. Ferry, H.Y. Lem, D.M. Hayes: J. Geophys. Res. *84*, 733-743 (1979)
2.32 S.C. Mossop: Nature London *199*, 325-326 (1963)
2.33 J.L. Gras: Geophys. Res. Lett. *3*, 533-536 (1976)
2.34 P.W. Hodge, F.W. Wright: Smithson Contrib. Astrophys. *5* (1962)
2.35 A.L. Lazrus, B. Gandrud, R.D. Cadle: J. Geophys. Res. *76*, 8083-8088 (1971)
2.36 A.L. Lazrus, B.W. Gandrud, R.N. Woodard, W.A. Cedlacek: Geophys. Res. Lett. *2*, 439-441 (1975)
2.37 A.L. Lazrus, E. Lorange, J.P. Lodge, Jr.: "New Automated Techniques for Sulfate Ion and for Total Inorganic Fixed Nitrogen", in *Trace Inorganics in Water*, ed. by R.F. Gould (Amer. Chem. Soc. Publ., Washington, D.C. 1968) pp.164-171
2.38 R.D. Cadle: Trans. AGU (EOS) *53*, 812-820 (1972)
2.39 A.L. Lazrus, B.W. Gandrud, R.N. Woodard, W.A. Sedlacek: J. Geophys. Res. *81*, 1067-1070 (1976)
2.40 N.H. Farlow: Anal. Chem. *29*, 883-885 (1957)
2.41 E.K. Bigg, A. Ono, J.A. Williams: Atmos. Environ. *8*, 1-13 (1974)
2.42 N.H. Farlow, K.G. Snetsinger, D.M. Hayes, H.Y. Lem, B.M. Tooper: J. Geophys. Res. *83*, 6207-6211 (1978)
2.43 N.H. Farlow, G.V. Ferry, H.Y. Lem: "Analysis of Individual Particles Collected From the Stratosphere", in *Space Research XIII* (Akademie, Berlin 1973) pp.1153-1157
2.44 E.S. Etz, G.J. Rosasco, J.J. Blaha: "Observation of the Raman Effect from Small, Single Particles; Its Use in the Chemical Identification of Airborne Particulates", in *Environmental Pollutants*, ed. by T.Y. Toribara, J.R. Coleman, B.E. Dahneke, I. Feldman (Plenum, New York 1978) pp.413-458
2.45 W.C. Cunningham, E.S. Etz, W.H. Zoller: "Raman Microbeam Characterization of South Pole Aerosol", in *Microbeam Analysis*, ed. by D. Newberry (San Francisco Press, San Francisco 1979) pp.148-153
2.46 J.L. Gras, C.G. Michael: J. Appl. Meteorol. *18*, 855-860 (1979)
2.47 H.A. Miranda, Jr., J. Dulchinos: "Balloon Measurements of Stratospheric Aerosol Size Distributions Following A Volcanic Dust Incursion"; Tech. Rpt. AFCRL 75-0518, Air Force Cambridge Research Laboratory (1975)
2.48 J.M. Rosen: J. Geophys. Res. *69*, 4673-4676 (1964)
2.49 C.E. Junge: *Air Chemistry and Radioactivity* (Academic, New York 1963) pp.115-118
2.50 K. Willeke, K.T. Whitby: J. Air Pollut. Control. *25*, 529-534 (1975)
2.51 V.R. Oberbeck, N.H. Farlow, G.V. Ferry, H.Y. Lem, D.M. Hayes: Geophys. Res. Lett. *8*, 18-20 (1981)
2.52 R.G. Pinnick, J.M. Rosen, D.J. Hofmann: J. Atmos. Sci. *33*, 304-314 (1976)
2.53 D.M. Hunten, R.P. Turco, O.B. Toon: J. Atmos. Sci. 1342-1357 (1980)
2.54 J.M. Rosen, D.J. Hofmann, K.H. Kaselau: J. Appl. Meteorol. *17*, 1737-1740 (1978)
2.55 *U.S. Standard Atmosphere, 1976*, NOAA-S/T 76-1562 (U.S. Govt. Print. Office, Washington D.C. 1976) pp.227
2.56 J.M. Rosen, D.J. Hofmann: Geophys. Res. Lett. *7*, 669-672 (1980)
2.57 J.L. Gras, J.E. Laby: J. Geophys. Res. *84*, 303-307 (1979)
2.58 O.B. Toon, N.H. Farlow: Annu. Rev. Earth Planet. Sci. *9*, 19-58 (1981)
2.59 O.B. Toon, J.B. Pollack: J. Appl. Meteorol. *15*, 225-246 (1976)
2.60 R.D. Cadle, G.W. Grams: Rev. Geophys. Space Sci. *13*, 475-501 (1975)
2.61 D. Hayes, K. Snetsinger, G. Ferry, V. Oberbeck, N. Farlow: Geophys. Res. Lett. *7*, 974-976 (1980)
2.62 J.M. Rosen: J. Appl. Meteorol. *10*, 1044-1046 (1971)
2.63 R.P. Turco, P. Hamill, O.B. Toon, R.C. Whitten, C.S. Kiang: J. Atmos. Sci. *36*, 699-717 (1979)

2.64 E.K. Bigg: J. Atmos. Sci. *33*, 1080-1086 (1976)
2.65 N.H. Farlow, V.R. Oberbeck, D.S. Colburn, G.V. Ferry, H.Y. Lem, D.H. Hayes: Geophys. Res. Lett. *8*, 15-17 (1981)
2.66 E.K. Bigg: J. Atmos. Sci. *32*, 910-917 (1975)
2.67 J.L. Gras: Nature London *271*, 231-232 (1978)
2.68 A.L. Lazrus, R.D. Cadle, B.W. Gandrud, J.P. Greenberg, B.J. Huebert, W.I. Rose, Jr.: J. Geophys. Res. *84*, 7869-7875 (1979)
2.69 B.W. Gandrud, A.L. Lazrus: Geophys. Res. Lett. *8*, 21-22 (1981)
2.70 N.H. Farlow, K.G. Snetsinger, V.R. Oberbeck, G.V. Ferry, G. Polkowski, D.M. Hayes: "Time Variations of Aerosols in the Stratosphere Following Mt. St. Helens Eruptions", in 1980 Symposium and Workshop on Mt. St. Helens Eruption Proc. (in press, 1981)
2.71 J.M. Rosen, D.J. Hofmann: "Dustsonde Measurements of the Mt. St. Helens Volcanic Dust Cloud over Wyoming", in 1980 Symposium and Workshop on Mt. St. Helens Eruption Proc. (in press, 1981)
2.72 G.S. Kent: "SAGE Measurements of Mt. St. Helens Volcanic Aerosols", in 1980 Symposium and Workshop on Mt. St. Helens Eruption Proc. (in press, 1981)
2.73 S.C. Mossop: Geochim. Cosmochim. Acta *29*, 201-207 (1965)
2.74 D.E. Brownlee: "Microparticle Studies by Sampling Techniques", in *Cosmic Dust*, ed. by J.A. McDonnell (Wiley, New York 1978) Chap.5
2.75 R. Ganapathy, D.E. Brownlee: Sciene *206*, 1075-1077 (1979)
2.76 L. Janos: Science 80, *1*, 44-55 (1980)
2.77 D.E. Brownlee, G.V. Ferry, K. Tomandi: Science *191*, 1270-1271 (1976)
2.78 S.C. Mossop: Nature London *203*, 824-827 (1964)
2.79 N.H. Farlow, V.R. Oberbeck, K.G. Snetsinger, G.V. Ferry, G. Polkowski, D.M. Hayes: Science *211*, 832-834 (1981)
2.80 D.J. Hofmann, J.M. Rosen, T.J. Pepin, R.G. Pinnick: J. Atmos. Sci. *32*, 1446-1456 (1975)
2.81 A.L. Lazrus, B.W. Gandrud: Geophys. Res. Lett. *4*, 521-522 (1977)
2.82 D.J. Hofmann, J.M. Rosen, J.M. Kiernan: J. Atmos. Sci. *33*, 1782-1788 (1976)
2.83 D.J. Hofmann, J.M. Rosen: Science *208*, 1368-1370 (1980)
2.84 J.M. Rosen, D.J. Hofmann, J. Laby: J. Atmos. Sci. *32*, 1457-1462 (1975)
2.85 H.A. Miranda, Jr., J. Dulchinos, H.P. Miranda: "Stratospheric Balloon Aerosol Particle Counter Measurements"; Tech. Rpt. AFCRL 73-0700, Air Force Cambridge Research Laboratory (1973)
2.86 E.D. Hinkley (ed.): *Laser Monitoring of the Atmosphere*, Topics in Applied Physics, Vol.14 (Springer, Berlin, Heidelberg, New York 1976)
2.87 M.P. McCormick, W.H. Fuller, Jr.: AIAA J. *11*, 244-246 (1973)
2.88 P.B. Russell, W. Viezee, R.D. Hake, Jr., R.T.H. Collis: J.R. Meteorol. Soc. *102*, 619-639
2.89 R. Reiter, H. Jager, W. Carnuth, W. Funk: Arch. Meteorol. Geophys. Bioklimatol. Ser. B *27*, 121 (1979)
2.90 M. Hirono, M. Fujiwara, O. Uchino, T. Itabe: Can. J. Chem. *52*, 1560-1568 (1974)
2.91 R.J. Fox, G.W. Grams, B.G. Schuster, J.A. Weinman: J. Geophys. Res. *78*, 7789-7801
2.92 F.G. Fernald, B.G. Schuster, E.F. Danielsen, D.G. Deaven: Opt. Qantum. Electron. *7*, 141-145 (1975)
2.93 W.H. Fuller, Jr., W.H. Hunt, B.R. Rouse: "Stratospheric Aerosol Measurements with a 1.06 μm Airborne Lidar", *Conf. Abstrs. 10th Int. Laser Radar Conf.*, Silver Spring, MD, USA, October 6-9, 1980 (Amer. Meteor. Soc., Boston 1980)
2.94 Shuttle Atmospheric Lidar Research Program: *Final Report of the Atmospheric Lidar Working Group.* NASA Spec. Publ. SP-433 (1979)
2.95 P.B. Russell, B.M. Morley, J.M. Livingston, G.W. Grams, E.M. Patterson: "Improved Simulation of Aerosol, Cloud, and Density Measurements by Shuttle Lidar"; Final Rpt. 1215, SRI International Menlo Park, CA, Contract NAS1-16052, to NASA Langley Research Center, Hampton, VA (1981)
2.96 P.B. Russell, T.J. Swissler, M.P. McCormick: Appl. Opt. *22*, 3783-3797 (1979)
2.97 G. Fiocco, L.D. Smullin: Nature London *199*, 1275-1276 (1963)
2.98 G. Fiocco, G.W. Grams: J. Atmos. Sci. *21*, 323-324 (1964)
2.99 M.G.H. Ligda: "Meteorological Observations with Lidar", in *1965 World Conference on Meteorology* (Amer. Meteor. Soc., Boston 1965) pp.482-489

2.100 G.W. Grams, G. Fiocco: J. Geophys. Res. 72, 3523-3542 (1967)
2.101 R.T.H. Collis, M.G.H. Ligda: J. Atmos. Sci. 23, 255-257 (1966)
2.102 B.R. Clemesha, G.S. Kent, R.W.H. Wright: Nature London 237, 328-329 (1966)
2.103 M.P. McCormick: "The Use of Lidar for Stratospheric Measurements", NASA Tech. Mem. 74085 (1977)
2.104 P.B. Russell, R.D. Hake, Jr.: J. Atmos. Sci. 34, 163-177 (1977)
2.105 D.J. Gambling, K. Bartusek, W.G. Elford: J. Atmos. Terr. Phys. 33, 1403-1413 (1971);
See also S.A. Young, W.G. Elford: "Laser Observations of Stratospheric Aerosols at Adelaide (35°S), 1969-1973"; Inter. Rep. ADP 119, Dept. of Physics, University of Adelaide, Australia (1975)
2.106 B.R. Clemesha, S.N. Rodrigues: J. Atmos. Terr. Phys. 32, 1119-1124 (1971)
2.107 B.G. Schuster: J. Geophys. Res. 75, 3123-3132 (1970)
2.108 M.T. Ottway: "Laser Radar Observations of the 20-km Aerosol Layer", 4th Conf. Laser Atmospheric Studies, Tucson, AZ (1972)
2.109 J.B. Pollack, O.B. Toon, C. Sagan, A. Summers, B. Baldwin, W. Van Camp: J. Geophys. Res. 81, 1071-1083 (1976)
2.110 G.B. Northam, J.M. Rosen, S.H. Melfi, T.J. Pepin, M.P. McCormick, D.J. Hofmann, W.H. Fuller, Jr.: Appl. Opt. 13, 3216-3421 (1974)
2.111 F.G. Fernald, B.G. Schuster: J. Geophys. Res. 82, 433-437 (1977)
2.112 M.P. McCormick, W.H. Fuller, Jr.: Appl. Opt. 14, 4 (1975)
2.113 E.E. Remsberg, G.B. Northam: Trans. Am. Geophys. Union 56, 365 (1975)
2.114 R.W. Fegley, H.T. Ellis: Geophys. Res. Lett. 2, 139 (1975)
2.115 F.G. Fernald, C.L. Frush: Trans. Am. Geophys. Union 56, 366 (1975)
2.116 M. Fujiwara, T. Itabe, M. Hirono: Rep. Ionos. Space Res. Jpn. 29, 74-78 (1975);
See also M. Hirono, M. Fujiwara, T. Itabe, C. Nagasawa: J. Geomagn. Geoelectr. 29, 541-556 (1977)
2.117 B.R. Clemesha, D.M. Simonich: J. Geophys. Res. 83, 2403-2408 (1978)
2.118 M.P. McCormick, T.J. Swissler, W.P. Chu, W.H. Fuller, Jr.: J. Atmos. Sci. 35, 1296-1303 (1978)
2.119 T.J. Swissler, P. Hamill, M. Osborn, P.B. Russell, M.P. McCormick: J. Atmos. Sci. (to be published,1982)
2.120 D.J. Hofmann, J.M. Rosen: J. Atmos. Sci. 38, 168-181 (1981)
2.121 R.W. Fegley, H.T. Ellis, J.L. Hefter: J. Appl. Meteorol. 19, 683-690 (1980)
2.122 S. West: Sci. News 115, 314-318 (1979)
2.123 E.E. Remsberg, E.V. Browell, G.B. Northam: Bull. Am. Meteorol. Soc. 57, 1152-1153 (1976)
2.124 M.P. McCormick: "Lidar Measurements of Mt. St. Helens Eruption"; in 1980 Symposium and Workshop on Mt. St. Helens Eruption Proc. (in press, 1981)
2.125 M.P. McCormick: "Report of the Panel on Remote Measurements of Mt. St. Helens Effluent"; in 1980 Symposium and Workshop on Mt. St. Helens Eruption Proc. (in press, 1981)
2.126 R.E. Newell, J.W. Kidson, D.G. Vincent, G.J. Boer: *The General Circulation of the Tropical Atmosphere*, I (MIT Press, Cambridge, MA 1972)
2.127 R. Reiter, H. Jaeger, W. Carnuth, W. Funk: Geophys. Res. Lett. 7, 1099-1111 (1980)
2.128 P.B. Russell, M.P. McCormick, T.J. Swissler, W.P. Chu, J.M. Livingston, W.H. Fuller, J.M. Rosen, D.J. Hofmann, L.R. McMaster, D.C. Woods, T.J. Pepin: J. Atmos. Sci. 38, 1295-1312 (1981)
2.129 P.B. Russell, T.J. Swissler, M.P. McCormick, W.P. Chu, J.M. Livingston, T.J. Pepin: J. Atmos. Sci. 38, 1279-1294 (1981)
2.130 J.B. Pollack, M.P. McCormick: Geophys. Res. Lett. 8, 1 (1981)
2.131 B.W. Gandrud, A.L. Lazrus: Geophys. Res. Lett. 8, 21-22 (1981)
2.132 N.H. Farlow, V.R. Oberbeck, D.S. Colburn, G.V. Ferry, H.Y. Lem, D.M. Hayes: Geophys. Res. Lett. 8, 15-17 (1981)
2.133 R.D. Cadle, C.S. Kiang, J.F. Louis: J. Geophys. Res. 81, 3125-3132 (1976)
2.134 R.D. Cadle, F.G. Fernald, C.L. Frush: J. Geophys. Res. 82, 1783-1786 (1977)
2.135 P.B. Russell, R.D. Hake, Jr., W. Viezee: *Proc. Symp. Radiation in the Atmosphere*, ed. by H.-J. Bolle (Science Press, Princeton, NJ 1977) pp.141-143
2.136 R.D. Cadle, G.W. Grams: Rev. Geophys. Space Phys. 13, 475-501 (1975)
2.137 J.B. Pollack, O.B. Toon, A Summers, W. Van Camp, B. Baldwin: J. Appl. Meteorol. 15, 247-258 (1976)

2.138 Harshvardon, R.D. Cess: Tellus *28*, 1-10 (1976)
2.139 M.P. McCormick, P. Hamill, T.J. Pepin, W.P. Chu, T.J. Swissler, L.R. McMaster: Bull. Am. Meteorol. Soc. *60*, 1038-1046 (1979)
2.140 A.E.S. Green, A. Deepak, B.J. Lipofsky: Appl. Opt. *10*, 1263-1279 (1979)
2.141 A. Deepak: In *Inversion Methods in Atmospheric Remote Sounding*, ed. by A. Deepak (Academic, New York 1977) pp.265-296
2.142 H.L. Malchow, C.K. Whitney: In *Inversion Methods in Atmospheric Remote Sounding*, ed. by A. Deepak (Academic, New York 1977) pp.217-263
2.143 W.P. Chu, M.P. McCormick: Appl. Opt. *18*, 1404-1413 (1979)
2.144 P.B. Russell, M.P. McCormick, L.R. McMaster, T.J. Pepin, W.P. Chu, T.J. Swissler: NASA Tech. Memo. *78747* (1978)
2.145 P.B. Russell (ed.): NASA Tech. Memo *80076* (1979) (Available from the NASA Langley Research Center Hampton, VA)
2.146 Guidelines for the Aerosol Climatic Effects Special Study: NASA Tech. Memo. *78554* (1979) (Available from the NASA Ames Research Center, Moffet Field, CA)
2.147 D.E. Miller, P.C. Simon, R. Fantechi: *European Ground-Truth Plan for SAGE* February 1979 (Available from NASA LaRC, MS/234, Hampton, VA, USA, 23665)
2.148 E.F. Harrison, R.N. Green, D.R. Brooks, G.F. Lawrence, and M.P. McCormick: "Mission Analysis for Satellite Measurements of Stratospheric Constituents by Solar Occultation", AIAA 13th Aerospace Science Meeting, Pasadena, CA (January, 1975)
2.149 M.P. McCormick: NASA Reference Publication 1081 (1981)
2.150 M.P. McCormick, W.P. Chu, G.W. Grams, P. Hamill, B.M. Herman, L.R. McMaster, T.J. Pepin, P.B. Russell, H.M. Steele, T.J. Swissler: Science *214*, 328-331 (1981)
2.151 P. Hamill, H.M. Steele, M.P. McCormick, W.P. Chu, T.J. Swissler: J. Atmos. Sci. (to be published, 1982)
2.152 M.P. McCormick, H.M. Steele, P. Hamill, W.P. Chu, T.J. Swissler: J. Atmos. Sci. (to be published, 1982)
2.153 M.P. McCormick, C.S. Kent, G.K. Yue, D.M. Cunnold: Science (to be published, 1981)
2.154 R.E. Newell (ed.): *Mt. St. Helens Eruption of 1980*, Atmospheric Effects and Potential Climatic Impact (to be published, 1982)
2.155 R.I. Tilling: *SPIE 81 EAST*, Proceedings Annual Technical Meeting, Vol.278 Washington, DC, USA, April 20-24, 1981 (to be published, 1981)
2.156 M.P. McCormick: Nature London *290*, 88 (1981)

Additional Reference

M.P. McCormick: NASA Technical Publication 83217 (1981)

3. The Chemical Kinetics of Aerosol Formation

R. G. Keesee and A. W. Castleman, Jr.

With 3 Figures

There are three general aspects to be considered in understanding the formation of the stratospheric aerosol. The first involves the chemical origins of the components that compose the aerosol. Precursor gaseous species can be transported to the stratosphere where chemical reactions may convert them to condensable species, while solid particles may also be directly injected into the stratosphere. Additionally, the possible formation of new aerosol particles by the nucleation processes of condensable species in the presence of primary (or injected) particles must be considered. Finally, account must also be taken of the evolving characteristics of the aerosol itself through subsequent growth, coagulation, and other attendant aging processes. The chemical composition after nucleation, or transport into the stratosphere, may be modified by heterogeneous reactions between aerosol particles and the surrounding gaseous constituents. In addition to heterogeneous reactions, the size distribution of the aerosol particles is also controlled by condensation or evaporation and physical processes such as coagulation and sedimentation.

This chapter considers the development of the stratospheric aerosol from the viewpoint of chemical kinetics. As anticipated from the above statements, the discussion is divided into three categories: chemical origins, nucleation mechanisms, and heterogeneous reactions and growth.

3.1 Chemical Origins

The pioneering studies of JUNGE and co-workers [3.1] established that a substantial component of the stratospheric aerosol layer is sulfate. This finding has led to extensive work to ascertain the origin and mechanisms of sulfate aerosol formation [3.2,3]. Although early measurements indicated that the sulfate ions were chemically combined with ammonium ions, later studies [3.4,5] suggested that this conclusion must be viewed with caution since handling procedures often introduce ammonia as contamination, which sometimes leads to the formation of ammonium sulfate by subsequent reactions *after* collection.

As a result of the work of LAZRUS and co-workers [3.6,7], these aerosols are also known to contain elements such as Si, Na, Cl, Mn, Br, and Ca. However, compositional analyses of the stratospheric aerosol by LAZRUS and GANDRUD [3.7] suggest that the quantity of cations (other than protons) is insufficient to chemically balance the sulfate. Consequently, it is generally assumed that sulfuric acid represents the major component of the stratospheric aerosol.

On the basis of thermodynamic considerations, TOON and POLLACK [3.8] concluded that sulfuric acid droplets should be stable at the temperature and the water vapor concentrations of the stratosphere. This conclusion follows from the fact that the partial pressures of both sulfuric acid vapor and water vapor in equilibrium with concentrated sulfuric acid solutions are very low at stratospheric temperatures. The authors estimate that the stratospheric droplets should be either a supercooled liquid or a solid composed of an approximately 75% sulfuric acid solution. Attempts to ascertain chemical composition have been made by ROSEN [3.9] in which the stratospheric aerosol was evaporated in situ, and its temperature of evaporation measured. Likewise, these findings indicate that a 75% (by mass) sulfuric acid solution is the major component of the aerosol. Recent morphologic studies of collected particles by HAYES et al. [3.5] also support this conclusion.

In early publications, JUNGE and his co-workers provided circumstantial evidence that the sulfate aerosols are not carried directly into the stratosphere but are most likely generated by gas-to-particle conversion reactions of SO_2 or H_2S imported into the stratosphere [3.1]. Subsequently, CASTLEMAN et al. [3.10] made a detailed study of the sulfur isotopic ratio of the stratospheric aerosol which led to similar conclusions. The observations of CASTLEMAN et al. have shown that during periods of large volcanic eruptions, sulfate aerosol is generated by an in situ oxidation mechanism rather than by simple input of tephra contained in the eruption clouds.

A clear correlation between sulfur concentration and volcanic activity was found from the data of CASTLEMAN et al. [3.10]. The perturbation to the total stratospheric dust burden by the paroxysmal eruption of Mt. Agung has been well documented and there was a large change in the sulfate concentration as well. The data show that the sulfate concentration in the southern hemisphere increased by approximately two orders of magnitude within a year after the eruptions but took some seven months or more to reach a peak concentration. Similar trends have been observed as a result of the eruptions of Fuego [3.11] and Mt. St. Helens [3.12,13]. Various sources of sulfur compounds are known to have specific ratios of ^{32}S to ^{34}S and this isotopic ratio can be used as an indication of the original source of the sulfur. A long-term systematically varying trend in the isotopic ratio following input of the reactive species is suggestive of subsequent chemical reactions. Furthermore, tephra is known to settle out relatively rapidly, and it must be concluded that the observed changes in isotopic ratio and concentrations are evidence for a gas-to-particle conversion mechanism.

Evidence for a gas-to-particle conversion mechanism based on the temporal variation of the isotopic ratios of sulfur is as follows. Characteristic trends are invariably observed following major volcanic events. An abrupt decline in the ^{34}S enrichment following the eruption of Mt. Agung and a much smaller decline in magnitude after the smaller eruption of Fernandina were observed [3.10]. After the passage of several years without another major eruption, the isotopic ratios always showed a tendency to return to preeruption values.

The data have been interpreted by recognizing that for short time intervals following major eruptions the additional sulfate contribution by all other processes is comparatively negligible, and the observed continual change in the sulfur isotopic composition for stratospheric aerosols is the result of fractionation processes. As a reaction proceeds to completion, the sulfate product would be expected to exhibit a unidirectional change in isotopic composition over a time span closely related to the eruption. Isotope fractionation is also observed with altitude. Since fractionation would not be expected on a purely physical basis after a few days following an eruption, these changes are clearly indicative of an in situ reaction.

Isotopic data also suggest a common source of sulfur in the stratosphere of both the northern and southern hemispheres. In the absence of major volcanic eruptions, the values are virtually the same. Yet, sulfur from anthropogenic sources has a different isotopic ratio than the ones generally observed. Furthermore, anthropogenic activities are far more extensive in the northern than the southern hemisphere, and interhemispherical mixing is not rapid in the upper atmosphere. Therefore, the findings indicate that the sulfate in the two hemispheres probably has a common origin during quiescent periods. One speculation is that biogenic sulfur released into tropical up-wellings may be this source. Most of the tropospheric sulfur compounds such as SO_2, H_2S, and dimethyl sulfide are believed to be sufficiently reactive to make it unlikely that these compounds would survive in the troposphere long enough to diffuse into the stratosphere in sufficient quantities [3.14]. Based on these considerations, CRUTZEN suggested that stratospheric aerosol formation in the absence of direct sulfur injection by volcanic eruptions may result from the diffusion of OCS into the stratosphere where it becomes photodissociated, eventually reacts to SO_2 and subsequently converts to sulfate aerosol.

Recently, INN et al. [3.15] have reported measurements of SO_2, OCS, and CS_2 abundances in the lower stratosphere. Their results show that OCS is the dominant species, offering further support for the suggestion of CRUTZEN. Carbon disulfide, on the other hand, appears to be at most only a minor contributor of stratospheric sulfur. It is apparent that during periods of high volcanic activity that sulfur is primarily introduced into the stratosphere by direct injection of SO_2; during quiescent periods atmospheric transport of OCS predominates. Therefore, an understanding of the chemical kinetics of aerosol formation requires knowledge about the transformation of gaseous OCS and SO_2 into particulate sulfate.

Table 3.1. OCS oxidation to SO_2 in the stratosphere

Reaction	Rate coefficient[a]
$OCS + h\nu \rightarrow S + CO$	$1.4 \times 10^{-9} \, s^{-1}$ @ 20 km
$OCS + O \rightarrow SO + CO$	$3.0 \times 10^{-11} e^{-2270/T} cm^3 s^{-1}$ [b]
$S + O_2 \rightarrow SO + O$	$2.2 \times 10^{-12} cm^3 s^{-1}$ [c]
$SO + O_2 \rightarrow SO_2 + O$	$3.0 \times 10^{-13} e^{-2800/T} cm^3 s^{-1}$
$SO + O_3 \rightarrow SO_2 + O_2$	$2.5 \times 10^{-12} e^{-1050/T} cm^3 s^{-1}$ [b]
$SO + NO_2 \rightarrow SO_2 + NO$	$1.4 \times 10^{-11} cm^3 s^{-1}$ [d]

[a] [3.16] and references therein
[b] See also [3.17]
[c] [3.18]
[d] [3.19]

The oxidation of OCS into SO_2 is reasonably well understood and the requisite rate coefficients have been experimentally determined. Table 3.1 lists the pertinent reactions and rate data. Since OCS is fairly unreactive, oxidation is believed to be largely initiated by the photolytic dissociation of OCS into sulfur atoms and CO. The sulfur can then be rapidly oxidized (on the order of a few seconds) to SO_2 by O_2, O_3, and NO_2.

Unfortunately, the details of the oxidation of SO_2 into sulfate are not well elucidated. FRIEND et al. [3.20] showed that the quantum yield for the photoexcitation of SO_2 is too low to make photooxidation of any consequence. Some ambiguity also results from uncertainties in stratospheric concentrations of likely oxidants. For instance, OH concentrations have not been measured in the lower stratosphere, and one must rely on atmospheric modeling calculations to assess the relative importance of various SO_2 oxidation reactions [3.21,22].

Laboratory determination of rate coefficients (Table 3.2) along with the results of modeling calculations of radical concentrations indicate that the dominant channel for SO_2 oxidation in the stratosphere is via the reaction with OH, probably first leading to the production of HSO_3. The reaction of SO_2 with CH_3O_2 is less important while oxidation by atomic oxygen and HO_2 are insignificant [3.22].

The overall gas-phase oxidation of SO_2 is relatively slow, having a time constant for SO_2 loss of around 10^7 s in the lower stratosphere. Therefore, this step is believed to be the rate limiting one. But one should note from collision kinetics and estimates of aerosol surface areas that a given gas molecule collides with an aerosol surface about every 10^4 to 10^5 s on the average. Therefore, heterogeneous reactions may be important. This aspect will be discussed later.

The fate of HSO_3, which is the product of the dominant SO_2 oxidation path, is not known. A simple reaction involving proton abstraction by OH to produce SO_3 has been proposed [3.6]. If it forms, the final stages of H_2SO_4 formation probably proceed through to rapid combination with water to form an adduct $SO_3 \cdot H_2O$ [3.27]. The adduct then presumably rearranges to sulfuric acid [3.29]. Other suggestions consider various clustering or heterogeneous reactions for HSO_3 [3.30,31]. In any event, this step is not thought to be rate limiting in the overall oxidation.

Table 3.2. Homogeneous SO_2 oxidation

Reaction	Rate coefficient	Reference
$SO_2 + OH + M \rightarrow HSO_3 + M$	$8.2 \times 10^{-13}/(7.0 \times 10^{17} + [M])$ $cm^6 s^{-1}$	[3.23]
$HSO_3 + OH \rightarrow SO_3 + H_2O$	1×10^{-11} $cm^3 s^{-1}$ (proposed rate)	[3.16]
$SO_2 + O + M \rightarrow SO_3 + M$	$3.4 \times 10^{-32} e^{-1130/T}$ $cm^6 s^{-1}$	[3.24]
$SO_2 + HO_2 \rightarrow SO_3 + OH$	$<1.0 \times 10^{-18}$ $cm^3 s^{-1}$	[3.25]
$SO_2 + CH_3O_2 \rightarrow SO_3 + CH_3O$	$\sim 6 \times 10^{-15}$ $cm^3 s^{-1}$	[3.26]
$SO_3 + H_2O \rightarrow H_2SO_4$	9.1×10^{-13} $cm^3 s^{-1}$	[3.27]
$HSO_3 + O_2 + M \rightarrow HSO_5 + M$ $HSO_5 + NO \rightarrow HSO_4 + NO_2$ $HSO_4 + HO_2 \rightarrow H_2SO_4 + O_2$	proposed	[3.28]

From the above discussion, it is apparent that the mechanism by which particulate sulfate is formed in the stratosphere is an open question. The incorporation of sulfate into the stratospheric aerosol may proceed via two different scenarios. First, homogeneous reactions may directly produce gas-phase H_2SO_4 and create a supersaturation of sulfuric acid with resultant nucleation and condensation. Second, the oxidation may be completed via heterogeneous reactions with the existing aerosol. In this case, H_2SO_4 vapor could result via evaporation from aerosol particles.

At present, the only measurement of stratospheric H_2SO_4 vapor concentration has been inferred from ion compositions around 36 km [3.32]. This altitude is somewhat above the stratospheric aerosol layer. One further complication in deciding if H_2SO_4 is actually supersaturated in the stratosphere is that the vapor pressure of H_2SO_4 at stratospheric conditions has not been resolved. In fact, recent measurements by ROEDEL [3.33] and AYERS et al. [3.34] are about an order of magnitude lower than the previously accepted value which was based on thermodynamic calculations. Realistically, the effect on H_2SO_4 vapor pressure of various impurities that are undoubtedly present in aerosol particles also should be considered.

3.2 Nucleation

The present section on nucleation considers the chemical kinetics of aerosol formation assuming that a supersaturation of sulfuric acid is produced via homogeneous reactions. An aerosol is formed when a gas-phase species undergoes a phase transformation to condensed phase particles. This process may proceed via a nucleation mechanism involving a) stepwise clustering of gas-phase molecules or b) interaction between small molecular clusters of the aerosol-forming species. The development of an aerosol and its size distribution can in principle be completely described by a generalized aerosol continuity equation (see Chap.4 and references therein) which describes the formation and loss processes for the spectrum of particles ranging from gas molecules, through small molecular clusters, to large aerosol particles. However, the complete kinetics required to solve this equation are not generally known, especially since the formation and interaction of molecular clusters containing tens of molecules would have to, in principle, be considered. Therefore, the formation of an aerosol particle is often treated as a problem separate from other processes such as condensation and coagulation.

Although an aerosol particle is composed of discrete chemical units, it is impractical to tabulate a discrete size distribution of particles. Nevertheless, it is unrealistic to consider small molecular clusters in a continuous manner since each chemical unit is a sizable portion of the cluster. The present section first considers formation processes for new aerosol particles as applicable to the stratosphere. The reactions involving aerosol particles themselves are considered in a subsequent section. Many of these general processes for aerosols are treated in several recent texts [3.35-37].

Nucleation processes may be classified according to the material basically responsible for inducing nucleation. These are a) homogeneous, b) heterogeneous, and c) heteromolecular. In the homogeneous case only those molecular species that constitute the condensing phase are involved in the nucleation process. Heterogeneous nucleation operates when these species condense onto the surface of a foreign body. In the heteromolecular case, a molecular entity acts as the foreign body which induces the nucleation of the condensable components.

A further distinction among the nucleation processes is made to denote the number of condensable substances which participate simultaneously to form new aerosol particles. Thus, nucleation may be unary, binary, ternary, etc. It should be noted that in the literature binary homogeneous nucleation is often referred to as heteromolecular nucleation, and so confusion with the above definition should be avoided.

The case of homogeneous nucleation has traditionally served as the basis for developing treatments of more complicated systems. We begin with its discussion since this provides a convenient way of defining certain terms which will be required later. Nucleation is assumed to evolve through the formation of molecular clusters or n-mers expressed by the following sequence of reactions

$$A + A + M \rightleftharpoons A_2 + M$$
$$A_2 + A + M \rightleftharpoons A_3 + M$$
$$A_{n-1} + A + M \rightleftharpoons A_n + M \tag{3.1}$$
$$A_n + A \rightleftharpoons A_{n+1}$$
$$A_{n-1}^* + A \rightleftharpoons A_n^*$$
$$A_n^* + A \rightarrow A_{n+1}^*$$

The abundance of monomer A is assumed to be much greater than that of all the clusters, so that mutual collisions of clusters can be neglected.

The foregoing reactions depict the fact that the early stages of n-mer formation require a third body M to remove from the n-mer the energy that is released in forming the cluster bond or, in the case of dissociation, to supply the energy to break off a clustering molecule. For a large cluster, the association energy becomes more efficiently distributed through its own internal rotational and vibrational motions and the lifetime of the complex becomes long compared to the mean time between collisions. The clustering kinetics is then effectively described as a two-body process independent of the total pressure. Finally, nucleation is considered to have occurred when a critical cluster A_n^* reacts with an additional molecule of A. At this point, further growth of this cluster is thermodynamically assured.

A complete description of nucleation requires information about the kinetics of each individual forward and reverse step. However, solving this large set of equations is impractical, and furthermore the details of the transition from 3-body to 2-body reaction kinetics is as yet an unresolved problem. Therefore, attempts to describe nucleation have focused on determining the population of critical clusters in a system. From the simple collision theory of a molecule colliding with a hard sphere (the critical cluster), the nucleation rate can be calculated. Also, a sticking coefficient may be introduced to account for the fact that not all collisions necessarily lead to the addition of a molecule.

Classical nucleation theory [3.35] assumes that a quasi-steady-state population of molecular clusters exists and that the rate of nucleation or aerosol formation is determined by the rate of collision of the monomeric gas molecules with critical clusters. The critical cluster is that cluster which corresponds to a free energy maximum during its growth starting from a free gas-phase molecule to a condensed-phase particle. The nucleation barrier $\Delta G^{\ddagger}$ is then the difference between this free energy maximum and the free energy minimum corresponding to the most stable subcritical cluster (most frequently the monomer itself in homogeneous nucleation). Thus, the nucleation rate J is expressed in terms of a Boltzmann-type relation, which determines the critical cluster concentration, given by

$$J = K \exp(-\Delta G^{\ddagger}/kT) \tag{3.2}$$

where k is the Boltzmann constant, T the absolute temperature, and the preexponential factor includes terms that express the collision dynamics of the monomer with the critical cluster, the concentration of species at the free energy minimum, and the deviation of the quasi-steady-state population of critical clusters from the constrained equilibrium one.

This formulation neglects any contribution of forming aggregates larger than the critical size by means of cluster-cluster collision. However, during heterogeneous or heteromolecular nucleation, a stable small cluster population may exist and cluster-cluster interactions should not be ignored. ZUREK and SCHIEVE [3.38] have shown that even for homogeneous nucleation cluster-cluster interactions can be important under some circumstances. Further approximation in the usual treatment arises because the scavenging of prenucleation embryos by any kind of preexisting aerosol is generally ignored. The effect of an existing aerosol on homogeneous nucleation has been studied theoretically only for the simplest case [3.39,40]. The assumption made is that no thermodynamic barrier to nucleation exists and that forward clustering rates, taken as hard-sphere collision rates, controls nucleation. GELBARD and SEINFELD [3.39] also considered the effect of allowing cluster-cluster agglomeration. Not surprisingly, in comparison to the classical Boltzmann distribution the loss of clusters onto aerosol particles depletes the number of critical clusters, while including cluster-cluster interactions enhances the population of larger clusters.

The demarcation between whether an aggregate of molecules is a gas-phase cluster or a condensed-phase aerosol particle, i.e., the point at which nucleation has occurred and growth commences, is not intuitively well defined. In nucleation theory this point is thermodynamically defined in terms of the critical cluster size, since further growth by condensation for particles larger than this size is assured. When no nucleation barrier exists, a critical cluster is not defined thermodynamically, and nucleation is limited only by the kinetics of the forward growth rates. In this case, no dividing line exists. In practice, the critically sized cluster is not directly observable, and the properties of nucleation are inferred from the aerosol particles that are observed. Therefore, the demarcation may be conveniently defined by a minimum detectable particle size. Computationally, the division may be chosen arbitrarily at some point where a "nucleation rate" is used as a boundary condition to represent the production of new particles of some minimum size.

When a nucleation barrier exists, the problem of determining the nucleation rate is largely one of calculating the requisite free energy changes for (3.2). Furthermore, this calculation is necessary to determine under what conditions the barrier disappears. The most broadly applicable approach is use of the capillarity

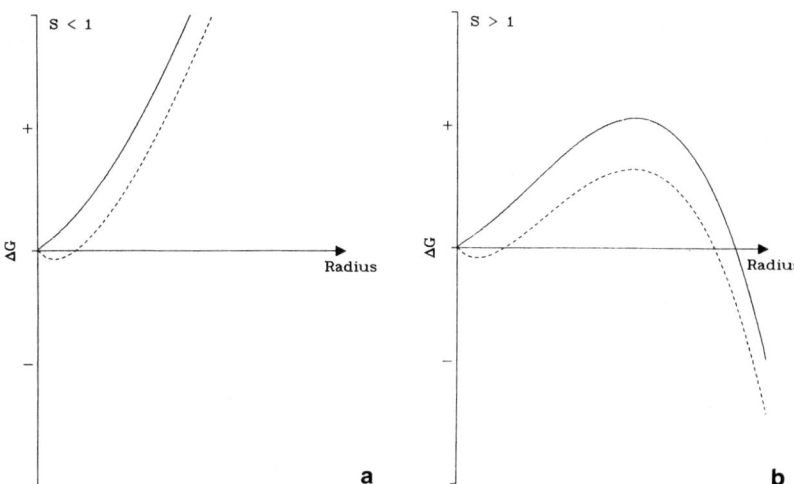

Fig.3.1a,b. The free energy change for forming a cluster or drop from gas-phase molecules as a function of the radius of the drop. The solid line is the usual homogeneous case and the dashed line depicts a case where a small stable cluster is formed for (a) the undersaturated case, and (b) the supersaturated case

approximation, often called the Kelvin or Thomson drop model, which is based on the properties of the bulk phase. In this model the free energy of forming a cluster of n molecules from the monomeric gas is due to condensation of n molecules, the formation of a surface, and effects due to a foreign nucleus if any. For the unary homogeneous case, and a spherical drop of n molecules, the free energy is expressed by the equation

$$\Delta G(n) = -nkT \ln S + 4\pi r^2 \sigma \tag{3.3}$$

where r is the radius (related to n by the density of the condensed medium) of the drop, and σ the surface tension. The saturation ratio S is defined as the partial pressure p of a component divided by its vapor pressure p_0 over the bulk condensed phase.

Figure 3.1 schematically shows the free energy changes of forming an agglomerate of n molecules from monomeric gas molecules. The solid lines represent the homogeneous cases and the dashed lines show cases in which stable small clusters exist. When the system is unsaturated (S<1) as in Fig.3.1a, the formation of the bulk phase is thermodynamically unfavorable. However, in some cases, especially in the heteromolecular or heterogeneous cases, a minimum in the free energy may occur so that a small stable cluster or droplet forms. In Fig.3.1b, the system is supersaturated (S>1), but a free energy maximum exists at the critical cluster size. Growth of molecular aggregates larger than the critical size reduces the free energy of the system and growth is, therefore, assured. On the other hand, to obtain the critical size requires surmounting the nucleation barrier $\Delta G^{\ddagger}$.

Whenever a system is supersaturated (S>1), the first term of (3.3) is negative and favors cluster growth. The second term is responsible for the existence of the nucleation barrier and is commonly referred to as the Kelvin effect. In other words, the surface energy results in a higher vapor pressure for a curved drop than for a flat surface of bulk. The Kelvin effect is such that the vapor pressure of a water drop of 0.1-μm radius is about 2% higher than that for a flat surface and about 25% higher for a 0.01-μm radius drop. Evidence in support of this effect and its validity has been reviewed by SKINNER and SAMBLES [3.41].

In nucleation theory, a macroscopic interpretation of a surface with a well-defined surface tension is at best questionable when one considers critical clusters of typically a few tens of molecules. However, the Thomson model remains popular because of its simplicity and its relative success in interpreting experimental results [3.42,43]. Another approach using macroscopic-type parameters is the Fisher droplet model [3.44].

Molecular models using statistical mechanics [3.45,46] to calculate cluster properties have a firmer theoretical footing and qualitatively produce results similar to macroscopic models. However, such models are not as yet amenable to problems of stratospheric nucleation because of the complexity and lack of the necessary information. Recent computer simulation techniques such as Monte Carlo [3.47] and molecular dynamics [3.48] methods circumvent the necessity of directly determining thermodynamic quantities. Using these methods ZUREK and SCHIEVE [3.49] have confirmed the concept of a critical cluster. As with molecular models, these methods have been limited to the treatment of only very elementary systems.

3.3 Nucleation Mechanisms in the Stratosphere

Assuming water and sulfuric acid as the primary components, the requirement that a gaseous constituent must be supersaturated with respect to a condensed phase in order to initiate the formation of an aerosol eliminates unary homogeneous nucleation as a viable mechanism for the creation of a stratospheric aerosol. The vapor pressure at stratospheric temperatures (around -50°C) of water and sulfuric acid over their pure solutions are well above the measured partial pressures of water and estimated sulfuric acid vapor concentrations [3.50]. More recently, the existence of low sulfuric acid concentrations in the stratosphere has been supported experimentally by ARNOLD and FABIAN [3.32]. Similarly, the unary heterogeneous or heteromolecular process could only occur if the vapor pressure is reduced by the foreign nucleus as, for instance, a "soluble" particle. However, a particle that may be soluble in the bulk phase may not act as a soluble nucleus as far as nucleation is concerned. A critical saturation must be exceeded in order that the soluble crystal absorb a sufficient amount of solvent molecules in order to dissolve. With water, for instance, the critical saturations are 0.55 (55% relative humidity) for sodium sulfate and 0.76 for sodium chloride [3.51]. The relative humidity of the stratosphere is typically below 1% so that such salts will behave as insoluble nuclei. Since water is much more abundant than sulfuric acid and the acid itself is hygroscopic, self-nucleation of sulfuric acid onto an acid-soluble crystal is improbable. Consequently, any discussion of nucleation in the natural stratosphere must include at least a binary process.

3.3.1 Binary Nucleation

As discussed in the foregoing section, binary nucleation is the most likely mechanism to lead to new particle formation in the stratosphere. When two pure substances are mixed, the vapor pressures of the individual components over the mixture may be lower than those over their pure states. Thus, a supersaturation in a mixed system may develop without the supersaturation of any single component with respect to its pure state. A suitable plot of the free energy versus composition would be as follows. A free energy surface could be pictured above a plane having one axis representing the number of molecules of one species in a cluster with the other axis being the number of molecules contributed by the other species. Thereupon, the critical cluster would be at the high point of the pass between the two insurmountable mountains lying along the axes.

In the Thomson formulation for the homogeneous case, the free energy of forming a drop of a composition defined by the number of molecules n_i of each component is given by the equation

$$\Delta G(n_1, n_2, \ldots n_\ell) = -kT \sum_{i=1}^{\ell} n_i \ln \frac{p_i}{p_{0i}} + 4\pi r^2 \sigma \qquad (3.4)$$

where p_i is the partial pressure of the ith component, and p_{0i} is its vapor pressure over a bulk solution of the composition determined by $(n_\ell, \ldots, n_\ell)$. The determination of the free energy barrier to nucleation at the critical size and composition is complicated by the fact that σ, p_{0i}, and the solution density to which r is related are implicit functions of the variables $n_\ell, \ldots, n_\ell$. YUE [3.52] concisely described the various numerical and graphical methods that have been applied to determine these properties for a binary system (where ℓ equals 2).

In the stratosphere the gaseous concentration of sulfuric acid molecules is much less than that of water. Therefore, a cluster or aerosol particle is quickly equilibrated with respect to water vapor, and the clustering and growth is determined by the collision rate of sulfuric acid molecules with these clusters and aerosol particles [3.53]. Consequently, the ratio p_i/p_{0i} for water may always be assumed to be unity.

3.3.2 Ternary Nucleation

The preceding discussion has considered only the simultaneous nucleation of two components since sulfuric acid and water are thought to be the major condensable species in the stratosphere. However, the possibility that three components promote nucleation under stratospheric conditions has been suggested by KIANG and HAMILL [3.54] with nitric acid as the third component. KIANG et al. [3.55,56] subsequently extended nucleation theory to ternary systems and considered the nitric acid-sulfuric acid-water system in the stratosphere. Unfortunately, vapor pressure data of ternary systems, particularly at stratospheric temperatures, are largely lacking. The suggestion by KIANG and HAMILL [3.54] that nitric acid might be saturated in the stratosphere with respect to HNO_3-H_2SO_4-H_2O solutions containing a significant amount of nitric acid (about 15% by weight) was based on an admittedly crude extrapolation to stratospheric conditions using data available in the *International Critical Tables* [3.57]. These data are from several different sources and are evidently inconsistent when examined carefully. The low vapor pressures for 15% nitric acid as estimated by KIANG and HAMILL appear to result from this inconsistency.

At present no evidence exists for nitric acid participation in stratospheric aerosol formation [3.7]. One should note, however, that compositional analysis of aerosols may also be misleading in distinguishing between the possible nucleation processes. The composition of stratospheric aerosols has been routinely determined at room temperature, where a component may become relatively more volatile than at stratospheric temperatures. On the other hand, a ternary mixture may develop subsequent to nucleation due to the reaction of aerosol particles with gaseous species. The possibility of ternary nucleation cannot as yet be discounted until more adequate data such as partial pressures of gaseous species over mixed systems become more readily available.

3.3.3 Binary Heterogeneous Nucleation

HAMILL et al. [3.50] have shown that homogeneous processes are very improbable due to the general availability in the stratosphere of foreign nuclei, along with preexisting aerosol particles. The nuclei include particles transported into the stratosphere from tropospheric, extraterrestrial, or volcanic sources [3.58]. Nuclei created in situ are generally of molecular dimensions such as ions and, therefore, may create aerosol particles via heteromolecular nucleation which is discussed in the next section.

Usually nuclei of tropospheric, volcanic, or extraterrestrial origin are large in comparison to the critical size of homogeneous nucleation. These nuclei would actually be aerosol particles in their own right. Heterogeneous nucleation as applied to the stratosphere may be considered a conversion of a nonsulfate aerosol into a sulfate-containing aerosol. The heterogeneous process on insoluble nuclei, which act only as sites for the sticking of the nucleating phase, modifies the surface energy required to reach the critical size. For instance, for a completely "wettable" nucleus of radius r_0, the surface free energy of forming a drop of radius r is given by $4\pi(r^2 - r_0^2)\sigma$. This results because the bare nucleus plus the monomeric gas molecules are the reference state of the system. If r_0 is larger than the critical radius, then no nucleation barrier would exist, and nucleation would be kinetically controlled by the time required to "wet" the nucleus.

For the case of a partially wettable nucleus, the macroscopic concept of a contact angle φ is often employed [Ref.3.35, Chap.12]. For a completely wettable nucleus the contact angle is $0°$, whereas for a completely nonnucleating unwettable nucleus φ would be $180°$. Basically, the contact angle determines the curvature, and consequently the surface energy, of the nucleating medium on a nucleus of a given shape or radius. A soluble nucleus would affect the vapor pressure in the system and act as a completely wettable nucleus. The effect of dissolved salts on the vapor pressure of H_2SO_4 over aqueous sulfuric acid is not known.

Nucleation on completely wettable insoluble nuclei has been calculated to be a very effective mechanism [3.50]. However, the problem of the wettability of these nuclei has not been adequately addressed, and FARLOW et al. [3.59] found that undissolved granules are present in only about one-third of the stratospheric aerosol particles during periods of low volcanic activity. In addition, the number of such nuclei in the stratosphere is still uncertain.

3.3.4 Heteromolecular Nucleation

a) *Ions*

Ion-induced nucleation is known to happen preferentially to homogeneous nucleation [3.60]. The electrostatic forces between ions and polar molecules leads to a distribution of stable ion-molecule complexes even in the absence of supersaturation and in some cases even covalent bonding is involved [3.61]. In the Thomson formulation (3.3), an ion is accounted for by an additional term $(q^2/2)(1 - 1/\varepsilon)(1/r - 1/r_i)$ which represents the energy for the creation of a cavity of charge q and radius r_i in a continuous medium of radius r with a dielectric constant ε. The continuous ionization of the stratosphere [3.62] provides a significant source of ions. If the lifetime against neutralization for the clustering ion is sufficiently long compared to the time required to induce nucleation, this mechanism may be operative. The average lifetime of an ion in the stratosphere is several hundred seconds, but the low sulfuric acid concentration implies slow H_2SO_4 clustering rates onto ions. Therefore, clustering kinetics, instead of thermodynamics as assumed in classical nucleation theory, may limit ion-induced nucleation in the stratosphere. In fact, it is the apparent comparability of ion lifetime with H_2SO_4 clustering rates that allowed ARNOLD and FABIAN [3.32] to calculate an H_2SO_4 concentration from ion cluster abundances.

Nevertheless, CASTLEMAN and TANG [3.63] established that small ion clusters actually represent a segment of the overall size distribution of atmospheric species. CHAN and MOHNEN [3.64] estimated a stable ion cluster distribution and relative critical cluster population using the Thomson drop model for the binary sulfuric acid-water system. The Thomson model is a very crude assumption in that specific ion-neutral interactions are not accounted for. Evidence now exists that stratospheric negative ions are mixed clusters of acids such as nitric acid, sulfuric acid, and possibly hydrochloric acid [3.65], while positive cluster ions contain water and an as yet unidentified species (possibly methyl cyanide [3.66]). The rather specific interactions of ions of opposite charge invalidate the use of the simple Thomson theory to predict the type of cluster ions present in the stratosphere.

Since ions form small stable clusters at a concentration of a few thousand per cubic centimeter, an attractive hypothesis is that ion cluster-cluster interactions may be important in producing nuclei upon which condensation can eventually occur. MOHNEN [3.67] first suggested this possibility. FERGUSON (private communication) subsequently pointed out that although the neutralization of oppositely charged ions may be exothermic, clustering may cause the charge neutralization of two ions to be endothermic so that their combination could produce a stable solvated ion pair. This is due to the fact that clustering stabilizes an ion by reducing the effective electron affinity of positive ions and increasing the ionization potential of negative ones. One should note that the products of ion-ion collisions and the stability of many of the feasible ion-neutral association clusters in the stratosphere are not known. Consequently, the degree of clustering required for a pair of ions to form a solvated ion pair in contrast to dissociated neutral products can at present only be qualitatively estimated.

Since an ion pair has a large dipole moment, ARNOLD [3.68] has suggested that further clustering of these ion pairs with ions to create multiion complexes may effectively promote nucleation by producing small electrolytic droplets as nuclei upon which condensation can occur. The recent detection of unexpectedly heavy positive ions below 30 km has been interpreted to support this suggestion [3.69]. In this case, of course, classical nucleation theory is quite inadequate for describing the rate of phase transformation and, therefore, aerosol formation.

b) *Radicals*

The reaction $OH + SO_2 + M \rightarrow HSO_3 + M$ is recognized as an important loss mechanism of sulfur dioxide [3.70]. The formation of aerosols initiated by this reaction in the presence of water (from which OH is formed by photolysis) is also well established [3.30]. NIKI et al. [3.71] have spectroscopically shown that the products appear to be liquid sulfuric acid containing varying amounts of water.

In these experiments, the role of radicals may be twofold. The products of radical reactions may create a supersaturated state of a substance with a low vapor pressure (e.g., H_2SO_4). Homogeneous nucleation would then ensue. Radicals may also act as nuclei to induce nucleation. Consequently, the role of the HSO_3 radical in forming the observed aerosol is not understood since the reactivity of HSO_3 has not been carefully studied and no quantitative treatment of radical-induced nucleation exists.

FRIEND et al. [3.20] have demonstrated that the products of the photolysis of H_2O-SO_2-O_2 mixtures rapidly produce droplets when introduced into a condensation nuclei counter in which the count rate (related to the nucleation rate) is dependent on the initial relative humidity of the preceding reaction zone. Due to the nature of this dependence, FRIEND et al. [3.31] have suggested that single sulfur molecules such as $H_2S_2O_6$, $H_2S_2O_8$, SO_3, or H_2SO_4, produced by the reaction or combination of free radicals, act as nuclei analogous to ions in ion-induced nucleation in the highly supersaturated condensation counter. ALLEN and KASSNER [3.72] have also suggested that H_2O_2 may promote the nucleation of water vapor. More work is definitely needed to test and quantify these hypotheses.

DAVIS et al. [3.30] have estimated that the dominant initial reaction path for HSO_3 in the stratosphere should be association with O_2. Association of water molecules to these radicals may also occur. The number of associated water molecules will be dependent on the relative humidity analogous to the cases for H_2SO_4 [3.73] or ions [3.63].

In regard to the stratosphere, DAVIS et al. [3.30] have suggested that reactions of sulfur radicals with radicals such as NO_2 may also be competitive with OH. Interestingly, species such as $NOHSO_4$ [3.74] and $(NH_4)_2S_2O_8$ [3.2] have been tentatively identified in samples of stratospheric aerosols. Such compounds as $H_2S_2O_6$, $H_2S_2O_8$, $NOHSO_4$ are also known to decompose or hydrolyze to sulfuric acid in acidic solutions, for example, [Ref.3.75, p.825]

$$H_2S_2O_8 + H_2O \rightarrow H_2SO_4 + H_2SO_5 \tag{3.5}$$

$$H_2SO_5 + H_2O \rightarrow H_2SO_4 + H_2O_2 \; . \tag{3.6}$$

These compounds present the possibility that all the HSO_3 may not lead to gaseous H_2SO_4, but instead produce gaseous $NOHSO_4$ or $H_2S_2O_8$ followed by hydrolysis in the aerosol condensed phase. The stability of these compounds against photodissociation is not known, but BENSON [3.76] has estimated them to be thermally stable under stratospheric conditions.

3.4 Growth and Heterogeneous Reactions

Nucleation is completed when a large nucleus becomes wetted or when a small one has reached critical size. At that point, these new aerosol particles continue to grow by coagulation, scavenging, condensation, or heterogeneous gas-aerosol reactions. The reverse process, namely, evaporation must also be considered.

In the stratosphere, the coagulative process due to Brownian diffusion is of primary importance. Gravitational coagulation, which results from the falling of a larger particle at a net rate with respect to a smaller one is unimportant in the stratosphere [3.16]. Also coagulation due to turbulent motion is usually not relevant to stratospheric conditions.

The general treatment of Brownian coagulation for the stratospheric aerosol is complicated by the particle dynamics. Figure 3.2 shows that the particle size range of interest extends into a transition region between where neither free-molecular kinetic nor slip-flow diffusive kinetic motion are fully applicable [3.58]. Equations have been formulated to interpolate this region so as to approximate the particle dynamics over the entire range from free-molecular to diffusive motion [3.78].

Whether condensation or evaporation operates on an aerosol sulfate particle depends on the sulfuric acid vapor pressure at given conditions. An aerosol particle is generally assumed to be essentially always at equilibrium with the surrounding water vapor. Thus, at a given temperature, water vapor concentration determines the concentration of sulfuric acid in the aqueous sulfuric acid aerosol particle. This in turn specifies the sulfuric acid vapor pressure over the particle [3.8]. Evaporation of the particle occurs if the sulfuric acid partial pressure is less than its corresponding vapor pressure. Condensation occurs in the opposite case. During these processes, the water content of the aerosol is adjusted practically instantaneously on the time scale of sulfuric acid addition or loss. Whereas coagulation decreases the total number of aerosol particles, as well as shifts the size spectrum to larger particles, condensation has no direct effect on the total particle concentration. Many of the details of coagulation and condensation are discussed by HAMILL et al. [3.79].

The scavenging of molecular clusters by aerosol particles lies between the domain of coagulation and condensation. All three processes are necessary to describe the effect of an existing aerosol on the total size spectrum. The major significance of scavenging is its effect on the steady-state molecular cluster distributions and consequently on heteromolecular and homogeneous nucleation. The scavenging of prenucleation embryos has very little effect on the size distribution of the much larger aerosol particles. The decrease of nucleation rates by a preexisting aerosol represents a feedback mechanism such that a balance is created between the total aerosol surface area and the particle production rate required to maintain that aerosol against losses due to sedimentation of large particles. For the

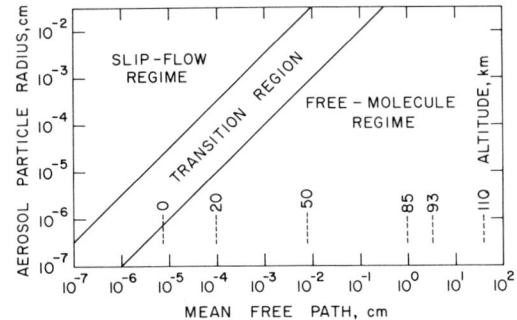

COAGULATION

$$\frac{\partial n(m,t)}{\partial t} = \frac{1}{2}\int_0^m K(m_1, m-m_1)n(m_1)n(m-m_1)dm_1$$

$$-n(m)\int_0^\infty K(m, m_1)n(m_1)dm_1$$

SETTLING VELOCITY

$$V_{SF} = \frac{2}{9}\frac{\rho g r^2}{\eta}A$$

$$V_{FM} = \frac{\rho g r}{(nMv)_{gas}}$$

Fig.3.2. Hydrodynamic characteristics of aerosol particles in the atmosphere. Approximate regions of the atmosphere (related to mean free path of gas molecules in terms of altitude) are noted for which specific sizes of aerosol particles are governed by free-molecule, transition, and slip-flow (continuum) mechanics. [Ref.3.77,p.153]

stratosphere, however, the contribution to the aerosol of particles transported from volcanic, tropospheric, and meteoric sources needs to be considered to establish the magnitude of in situ particle production. At present only rough estimates of these contributions exist [3.16,80,81].

Heterogeneous reactions may be controlled by any of these processes: 1) the transport of reacting species to the aerosol particle by free-molecular collision or diffusion, 2) by the surface reaction rate, or 3) by bulk reactions within the volume of the particle [3.36]. Each of these processes results in different growth laws. The radial growth rate (dr/dt) of a particle (assumed to be spherical) undergoing heterogeneous reactions controlled by transport is independent of the particle size if that particle is in the free-molecule regime (Fig.3.2). If the particle is in the slip-flow regime, the growth rate decreases as 1/r. For reactions controlled by a surface reaction rate, the radial growth is again independent of particle size, whereas for a bulk phase controlled reaction, the growth rate increases linearly with the radius of the particle. Growth by condensation/evaporation is controlled by transport. However, due to the dependence of vapor pressure on the curvature of the particle (the Kelvin effect discussed in Sect.3.2), the growth rate may be slower or even negative (evaporation) for smaller particles. The condensational radial growth rate as a function of particle radius for a pure substance is shown schematically in Fig.3.3. This figure depicts the Kelvin effect for small particles, the free-molecule regime for intermediately sized particles, and the diffusion-controlled regime for large particles. If a particle contains soluble nonvolatile impurities, then the growth law can be further complicated by the effect of the concentration of the impurities on the vapor pressure.

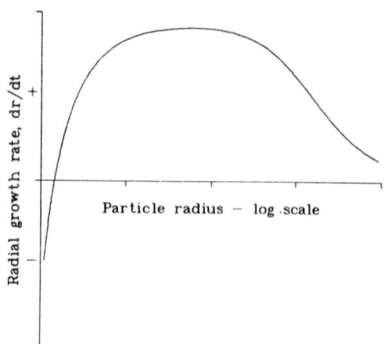

Fig.3.3. The dependence of the radial growth rate (for condensation) on the size of the particle

Heterogeneous reactions may have several consequences, not only with regard to the size distribution and chemical composition of the aerosol, but also to the composition of the surrounding atmosphere. These include the potential destruction of certain reactive intermediates such as free radicals which could be normally important in gas-phase reactions, catalytic influences on reactions between stable gas-phase constituents, or the stabilization of a product molecule in the condensed state which might otherwise readily dissociate in the gas phase.

Quantitatively, very little is known about catalytic reaction mechanisms or rates that might be important in the upper atmosphere. Therefore, in order to assess which mechanisms warrant attention, calculations were made where it was assumed that one species is adsorbed and others react with it upon each collision [3.82]. By comparing these rates to known homogeneous ones involving the same species, a rough prediction was made regarding what species might have potentially important surface-controlled chemistry. Admittedly this does not provide an unequivocal assessment of the importance, or lack thereof for the species under consideration, but it gives a way of screening and establishing priorities for laboratory investigation. Table 3.3 is a list of reactants, possible products, and an estimate of the maximum possible ratio of heterogeneous to homogeneous rates. From results for particle size distributions and elementary kinetic theory, a gas-phase molecule suffers a collision with an aerosol particle approximately every 10^4 to 10^5 s [3.81]. Therefore, for gaseous species that have homogeneous rates with time constants about 10^4 s or less, heterogeneous reactions will be a minor or negligible concern.

Based on the listing in Table 3.3, few of the systems warrant any attention whatsoever. A few which should be considered are the surface reactions between N_2O_5 and water, and O, NO, and NO_2 with O_3.

Since homogeneous oxidation of SO_2 is relatively slow, heterogeneous oxidation of SO_2 is a distinct possibility. Aqueous sulfuric acid solutions absorb SO_2 according to Henry's law [3.75] (Henry's constant K_H). The following reactions describe an uncatalyzed heterogenous oxidation mechanism for SO_2 [3.83]:

Table 3.3. Comparison of surface and gas-phase reaction rates[a]

Reactants	Products	Ratio, surface to gas-phase reaction rate
N_2O_5, H_2O	$2\ HNO_3$	$>10^3$
NO_2, O_3	NO_3, O_2	~ 10
CH_3O_2, NO	CH_3O, NO_2	~ 1
O_3, O	$2\ O_2$	200
NO, O_3	NO_2, O_2	~ 10
$HO_2\cdot$, O_3	$OH\cdot$, O_2	5.3×10^{-2}
$OH\cdot$, O_3	$HO_2\cdot$, O_2	1×10^{-3}
$CH_3O\cdot$, O_2	$CH_2O\cdot$, $HO_2\cdot$	1.6×10^{-5}
$Cl\cdot$, O_3	$ClO\cdot$, O_2	1×10^{-6}
$O\cdot$, O_2	O_3	2.5×10^{-7}
$CH_3\cdot$, O_2	$CH_3O_2\cdot$	5.8×10^{-8}
$H\cdot$, O_2	$HO_2\cdot$	1.7×10^{-10}

$$SO_2(\text{gas}) \overset{K}{\rightleftharpoons} SO_2 \cdot H_2O \ (\text{absorbed } SO_2)$$
$$SO_2 \cdot H_2O \overset{K_1}{\rightleftharpoons} H^+ + HSO_3^-$$
$$HSO_3^- \overset{K_2}{\rightleftharpoons} H^+ + SO_3^{2-} \tag{3.7}$$
$$SO_3^{2-} + \tfrac{1}{2} O_2 \overset{k}{\rightarrow} SO_4^{2-} \ .$$

The final step is generally agreed to be first order in SO_3^{2-} and independent of O_2 over a wide range of O_2 concentrations.

Because of the acid dissociation equilibria, the droplet concentration of SO_3^{2-} depends strongly on pH. Therefore, the droplet must absorb some buffering gas to maintain a sufficiently high pH to allow oxidation. No buffering mechanism is known for the stratospheric aerosol.

Alternatively, catalysts such as Mn^{2+} and Fe^{2+} promote SO_3^{2-} oxidation but the rate depends on both SO_3^{2-} and catalyst concentrations. Lacking any information about the quantity of Mn^{2+} or Fe^{2+} that might be available in the stratospheric aerosol, no evaluation of this mechanism can be made. Oxidation by H_2O_2 or O_3 may also be possible. However, BALDWIN and GOLDEN [3.84] observed no reaction between SO_2 and O_3 in the presence or absence of a sulfuric acid surface (> 95% H_2SO_4).

---

[a]Maximum surface reaction rate based on kinetic theory for collision rates, and assuming unit reaction probability upon collision

OLSYNA et al. [3.85] have shown that ozone destruction is not important on sulfuric acid surfaces. Therefore, this marginally important case may be omitted from further consideration unless substantial metalic oxide aerosols are found, in which case they may play a somewhat important role (FERGUSON [3.86] suggested the possible importance of ozone reactions with metal atoms which might be a source of oxide aerosols).

BALDWIN and GOLDEN [3.84] have investigated several heterogeneous reactions of gaseous constituents on a H_2SO_4 surface (<5% water by weight) at room temperature. Their results also showed that ozone is unreactive. However, for N_2O_5 they reported a lower limit to the collision reaction probability (i.e., the fraction of collisions that result in a chemical reaction) of about 4×10^{-5}. This limit, combined with the value given in Table 3.3, is inconclusive as to whether the heterogeneous reaction is important. Furthermore, one should note that BALDWIN and GOLDEN's results are not particularly applicable to the stratosphere since 1) the temperature dependences are not known, 2) the curvature of aerosol particles may effect surface properties, and 3) the aqueous sulfuric acid in the stratosphere is thought to be ≈25% water (by mass).

Some experiments have been performed using sulfuric acid aerosols. HUNTZICKER et al. [3.87] measured the reaction rate of NH_3 with sulfuric acid particles (diameter 0.3 to 1.4 μm). Their results indicated a slight increase of the reaction probability from about 0.15 for the smallest particles to about 0.3 for the largest. There was no observable effect by varying the humidity (and thus the H_2SO_4 weight fraction of the particles) over the range of 8% to 80%. Earlier measurements by CADLE and ROBBINS [3.88] found that the reaction probably was 0.1 for an essentially dry aerosol (98% H_2SO_4), and they could not measure the rate at 90% relative humidity and speculated that it must be near unity.

Recently, MARTIN et al. [3.89] have made measurements of the reactivity of Cl and ClO with aqueous sulfuric acid surface at conditions resembling the stratosphere, although they did use a film instead of an aerosol. In addition, the temperature dependence of the reaction probabilities was determined. However, the composition of the sulfuric acid-water film did vary systematically with temperature. The experimental results were used to predict the effect of these heterogeneous reactions on the stratospheric gas-phase Cl/HCl ratio. The ClO reaction was calculated to reduce this ratio by about 3% at 20 km. The Cl reaction had a negligible effect.

Stratospheric aerosols might possibly play a minor role in the production of HCl from the interaction of H_2SO_4 with NaCl [3.82]. Calculations show that the reaction is exothermic and the quantity of HCl so produced is likely to be limited by the quantity of NaCl, potentially from sea salt, which may be deposited in the stratosphere following major storms in equatorial regions. However, the kinetics of the reactions are not known under the low-temperature conditions prevailing in the stratosphere, and laboratory studies of this reaction are needed. The reaction is unlikely to be of major importance since, at most, only one HCl molecule could be produced per sulfur atom introduced into the stratosphere.

A consideration of the potential importance of stratospheric heterogeneous reactions on the behavior of the fluorocarbons has failed to provide any evidence in support of such a mechanism. AUSLOOS et al. [3.90] have demonstrated that unexpectedly large surface photochemical effects can result on certain surfaces, notably certain sands, leading to the destruction of some fluorocarbon compounds and N_2O. The mechanisms and significance of these catalytic processes are not well known, but even if they do occur, the quantity of potentially active surface material is far too low to be of importance in the stratosphere.

Finally, we consider the charge exchange between ions and aerosols and the charge balance of the aerosol. Since the mass of positive and negative ions are similar, the collision dynamics of the two with aerosol particles are similar. (Electrons are negligible in the lower stratosphere). As a result, the net charge of the aerosol should be nearly neutral. Studies [3.91] showed that, to a good approximation, aerosols maintain a Boltzmann charge distribution, i.e., the number of particles of radius r with charge Ze is related to the number of uncharged particles by the expression

$$N_z = N_0 \exp(-Z^2 e^2 / 2kTr) \quad . \tag{3.8}$$

More recently KOJIMA [3.92] has shown that the Boltzmann distribution underestimates the fraction of the aerosol which carries a charge for particles smaller than about 0.02 μm in radius. Most particles larger than 0.1 μm are at least singly charged. The rate of change of ion concentration, n, is given by

$$\frac{dn}{dt} = Q - \alpha n^2 - n(\beta A) \quad , \tag{3.9}$$

where t is time, Q ion production rate, α recombination coefficient of negative with positive ions, β the coefficient of ion attachment to aerosols, and A the aerosol surface area.

The stratosphere contains approximately 10^3 ion pairs/cm^3. Under stratospheric conditions, the recombination coefficient α is 10^{-7} to 10^{-6} cm^3/s from which it follows that the ratio of the loss of ions due to mutual recombination to that due to charge exchange upon collision with an aerosol particle is

$$\frac{\alpha n}{\beta A} \approx 10 \quad . \tag{3.10}$$

Therefore, the charge exchange processes with aerosols are of only secondary importance except when the aerosol concentration is increased following periods of intense volcanic activity. Ion-ion interactions resulting in the formation of multiion complexes could create heavier ions than expected from straightforward ion-molecule clustering. The presence of these heavier ions would add another loss mechanism for the lighter ions via light ion-multiion complex clustering.

3.5 Conclusions

Various sampling studies and numerical models have provided evidence that the in situ oxidation of sulfur-bearing gases [3.1,16] is responsible for the sulfate mass of the stratospheric aerosol. An extensive study of the temporal and spatial distribution of the sulfur isotope ratio by CASTLEMAN et al. [3.10] has borne this out. Using elementary modeling calculations and the results of laboratory experiments, CASTLEMAN et al. [3.70], DAVIS et al. [3.30], and MOORTGAT and JUNGE [3.23] have speculated that the stratospheric aerosol layer originates, at least in part, from SO_2 oxidation via OH. Another candidate for the origin of the sulfur component of the aerosol layer is COS. It is still unclear to what extent the oxidation of the sulfur species is completely homogeneous creating a supersaturation of H_2SO_4 with subsequent nucleation and condensation or heterogeneous with the ultimate sulfate production occurring directly on preexisting particles.

Very little is known concerning the origin of the primary small particles that form as a result of processes following the generation of the precursors to the prenucleation embryos. It is almost certain that homogeneous nucleation does not operate in the atmosphere and that the more relevant processes are those termed heteromolecular and heterogeneous nucleation. Furthermore, in the stratosphere processes that involve the interaction of more than one gaseous species participating in the formation of the condensed phase (particle), are certain to dominate under situations where new particle generation occurs.

The relative contribution of the various mechanisms proposed for the introduction of particles in the stratosphere has not been established. This problem is further complicated because the various nucleation processes and condensation compete for a limited supply of sulfur. Recently, HAMILL et al. [3.93] have considered the effect of the competition of several nucleation processes and condensation in model calculations. In light of the model results, their basic conclusion is that presently available observational techniques cannot unambiguously determine which nucleation mechanisms are responsible for the formation of stratospheric aerosol particles. Nevertheless, the general characteristics and extent of the stratospheric sulfate aerosol are reasonably well understood in terms of coagulation, condensation, evaporation, and sedimentation when a source for the generation of new particles is assumed.

References

3.1 C.E. Junge, J.E. Manson: J. Geophys. Res. *66*, 2163-2182 (1961)
3.2 J.P. Friend: Tellus *18*, 465-473 (1966)
3.3 J.P. Shedlovsky, S. Paisley: Tellus *18*, 499-503 (1966)
3.4 D. Klockow, B. Jablonski, R. Neissner: Atmos. Environ. *13*, 1665-1676 (1979)
3.5 D. Hayes, K. Snetsinger, G. Ferry, V. Oberbeck, N. Farlow: Geophys. Res. Lett. *7*, 974-976 (1980)

3.6 R.D. Cadle, A.L. Lazrus, W.H. Pollack: "Chemical Composition of Aerosol Particles in the Tropical Stratosphere," Proc. Symp. Tropical Meteorol., Honolulu, Hawaii, June 2-11, 1970, Section K-IV (1970) p.7
3.7 A.L. Lazrus, B.W. Gandrud: J. Geophys. Res. 79, 3424-3431 (1974)
3.8 O.B. Toon, J.B. Pollack: J. Geophys. Res. 78, 7051-7059 (1973)
3.9 J.M. Rosen: J. Appl. Meteorol. 10, 1044-1046 (1971)
3.10 A.W. Castleman, Jr., H.M. Munkelwitz, B. Manowitz: Tellus 26, 222-234 (1974)
3.11 D.J. Hofmann, J.M. Rosen: J. Geophys. Res. 82, 1435-1440 (1977)
3.12 R.L. Chuan, D.C. Woods, M.P. McCormick: Science 211, 830-832 (1981)
3.13 N.H. Farlow, V.R. Oberbeck, K.G. Snetsinger, G.V. Ferry, G. Polkowski, D.M. Hayes: Science 211, 832-834 (1981)
3.14 P. Crutzen: Geophys. Res. Lett. 3, 73-76 (1976)
3.15 E.C.Y. Inn, J.F. Vedder, D. O'Hara: Geophys. Res. Lett. 8, 13-25 (1981)
3.16 R.P. Turco, P. Hamill, O.B. Toon, R.C. Whitten, C.S. Kiang: J. Atmos. Sci. 36, 699-717 (1979)
3.17 J.S. Robertshaw, I.W.M. Smith: Int. J. Chem. Kinet. 12, 729-739 (1980)
3.18 M.A.A. Clyne, P.D. Whitefield: J. Chem. Soc. Faraday Trans. 2 75, 1327-1340 (1979)
3.19 M.A.A. Clyne, J. Macrobert: Int. J. Chem. Kinet. 12, 79-96 (1980)
3.20 J.P. Friend, R. Leifer, M.P. Trichon: J. Atmos. Sci. 30, 465-479 (1973)
3.21 A.B. Harker: J. Geophys. Res. 80, 3399-3401 (1975)
3.22 R.P. Turco, R.C. Whitten, O.B. Toon, E.C.Y. Inn, P. Hamill: J. Geophys. Res. 86, 1129-1139 (1980)
3.23 G.K. Moortgat, C.E. Junge: Pure Appl. Geophys. 115, 759-774 (1977)
3.24 D.D. Davis: Can. J. Chem. 52, 1405-1414 (1974)
3.25 R.A. Graham, A.M. Winer, R. Atkinson, J.N. Pitts: J. Phys. Chem. 83, 1563-1567 (1979)
3.26 C.S. Kan, R.D. McQuigg, M.R. Whitbeck, J.G. Calvert: Int. J. Chem. Kinet. 11, 921-933 (1979)
3.27 A.W. Castleman, Jr., R.E. Davis, H.R. Munkelwitz, I.N. Tang, W.P. Wood: Int. J. Chem. Kinet. Symp. No.1, 629-640 (1975)
3.28 D.D. Davis, G. Smith, G. Klauber: Science 186, 733-736 (1974)
3.29 P.M. Holland, A.W. Castleman, Jr.: Chem. Phys. Lett. 56, 511-514 (1978)
3.30 D.D. Davis, A.R. Ravishankara, S. Fisher: Geophys. Res. Lett. 6, 113-116 (1979)
3.31 J.P. Friend, R.A. Barnes, R.M. Vasta: J. Phys. Chem. 84, 2423-2436 (1980)
3.32 F. Arnold, R. Fabian: Nature London 283, 55-57 (1980)
3.33 W. Roedel: J. Aerosol Sci. 10, 375-386 (1979)
3.34 G.P. Ayers, R.W. Gillett, J.L. Gras: Geophys. Res. Lett. 7, 433-436 (1980)
3.35 H.R. Pruppacher, J.D. Klett: *Microphysics of Clouds and Precipitation* (Reidel, Dordrecht, Holland 1978) p.714
3.36 S.K. Friedlander: *Smoke, Dust, and Haze* (Wiley, New York 1977) p.317
3.37 S. Twomey: *Atmospheric Aerosols* (Elsevier, New York 1977) p.302
3.38 W.H. Zurek, W.C. Schieve: Phys. Lett. 67A, 42-45 (1978)
3.39 F. Gelbard, Seinfeld: J. Colloid Interface Sci. 68, 363-382 (1979)
3.40 P.H. McMurry, S.K. Friedlander: Atmos. Environ. 13, 1635-1651 (1979)
3.41 L.M. Skinner, J.R. Sambles: J. Aerosol. Sci. 3, 199-210 (1972)
3.42 H.L. Jaeger, E.J. Wilson, P.G. Hill, K.C. Russell: J. Chem. Phys. 51, 5380-5388 (1969)
3.43 A.W. Castleman, Jr., P.M. Holland, R.G. Keesee: J. Chem. Phys. 68, 1760-1767 (1978)
3.44 D. Stauffer, C.S. Kiang: Adv. Colloid Interface Sci. 7, 103-130 (1977)
3.45 B.N. Hale, P.L.M. Plummer: J. Atmos. Sci. 31, 1621-1651 (1973)
3.46 H.R. Hoare, P. Pal, P.P. Wegener: J. Colloid Interface Sci. 75, 126-137 (1980)
3.47 F.F. Abraham: J. Chem. Phys. 61, 1221-1222 (1974)
3.48 C.L. Briant, J.J. Burton: J. Chem. Phys. 63, 2045-2058 (1975)
3.49 W.H. Zurek, W.C. Schieve: J. Phys. Chem. 84, 1479-1482 (1980)
3.50 P. Hamill, C.S. Kiang, R.D. Cadle: J. Atmos. Sci. 34, 150-162 (1977)
3.51 F.E.M. O'Brien: J. Sci. Instrum. 25, 73-76 (1948)
3.52 G.K. Yue: J. Aerosol Sci. 10, 75-86 (1979)
3.53 P. Mirabel, J.L. Katz: J. Chem. Phys. 60, 1138-1144 (1974)
3.54 C.S. Kiang, P. Hamill: Nature London 250, 401-402 (1974)

3.55 C.S. Kiang, R.D. Cadle, P. Hamill, V.A. Mohnen: J. Aerosol Sci. *6*, 465-474 (1975)
3.56 C.S. Kiang, R.D. Cadle, G.K. Yue: Geophys. Res. Lett. *2*, 41-44 (1975)
3.57 F.C. Zeisberg: "Vapor Pressures, Boiling Points and Vapor Compositions for the System $H_2O-H_2SO_4-HNO_3$," in *International Critical Tables*, ed. by E.W. Washburn (McGraw-Hill, New York 1928) Chap.3, pp.306-308
3.58 A.W. Castleman, Jr.: Space Sci. Rev. *15*, 547-589 (1974)
3.59 N.H. Farlow, D.M. Hayes, H.Y. Lem: J. Geophys. Res. *82*, 4921-4929 (1977)
3.60 A.W. Castleman, Jr.: Adv. Colloid Interface Sci. *10*, 73-128 (1979)
3.61 R.G. Keesee, N. Lee, A.W. Castleman, Jr.: J. Chem. Phys. *73*, 2195-2202 (1980)
3.62 M. Nicolet: Planet. Space Sci. *23*, 637-649 (1975)
3.63 A.W. Castleman, Jr., I.N. Tang: J. Chem. Phys. *57*, 3629-3638 (1972)
3.64 L.Y. Chan, V.A. Mohnen: J. Aerosol Sci. *11*, 35-45 (1980)
3.65 F. Arnold, R. Fabian, E.E. Ferguson, W. Joos: Planet. Space Sci. *29*, 195-203 (1981)
3.66 E. Arijs, D. Nevejans, J. Ingels: Nature London *288*, 684-686 (1980)
3.67 V.A. Mohnen: "Discussion of the Formation of Major Positive and Negative Ions up to the 50-km Level", in *Mesospheric Models and Related Experiments*, ed. by G. Fiocco (Reidel, Dordrecht, Holland 1971) pp.210-219
3.68 F. Arnold: Nature London *284*, 610-611 (1980)
3.69 F. Arnold, G. Henschen: Geophys. Res. Lett. *8*, 83-86 (1981)
3.70 A.W. Castleman, Jr., I.N. Tang: J. Photochem. *6*, 349-354 (1976/77)
3.71 H. Niki, P.D. Maker, C.M. Savage, L.P. Breitenbach: J. Phys. Chem. *84*, 14-16 (1980)
3.72 L.B. Allen, J.L. Kassner: J. Colloid Interface Sci. *30*, 81-93 (1969)
3.73 R.H. Heist, H. Reiss: J. Chem. Phys. *61*, 574-581 (1974)
3.74 N.H. Farlow, K.G. Snetsinger, D.M. Hayes, H.Y. Lem: J. Geophys. Res. *83*, 6207-6211 (1978)
3.75 *Gmelin Handbuch der Anorganischen Chemie*, Schwefel, Teil B2 "Lieferung: Schwefelsauerstoffsäuren" (Springer, Berlin, Heidelberg, New York 1960)
3.76 S.W. Benson: Chem. Rev. *378*, 23-35 (1978)
3.77 G.M. Hidy, J.R. Brock: *Dynamics of Aerocolloidal Systems* (Pergamon, New York 1970)
3.78 N.A. Fuchs, A.G. Sutugin: "Highly Dispersed Aerosols", in *Topics in Current Aerosol Research*, ed. by G.M. Hidy, J.R. Brock (Pergamon, New York 1971) Chap.2
3.79 P. Hamill, O.B. Toon, C.S. Kiang: J. Atmos. Sci. *34*, 1104-1119 (1977)
3.80 D.M. Hunten, R.P. Turco, O.B. Toon: J. Atmos. Sci. *37*, 1342-1357 (1980)
3.81 R.P. Turco, O.B. Toon, P. Hamill, R.C. Whitten: J. Geophys. Res. *86*, 1113-1128 (1981)
3.82 A.W. Castleman, Jr., R.E. Davis, I.N. Tang, J.A. Bell: "Heterogeneous Process and the Chemistry of Aerosol Formation in the Upper Atmosphere", Proc. 4th Conf. Climate Impact Assess. Prog. Cambridge, MA, February 4-7, 1975, pp.470-477
3.83 S. Beilke, G. Gravenhorst: Atmos. Environ. *12*, 231-239 (1978)
3.84 A.C. Baldwin, D.M. Golden: Science *205*, 562-563 (1979)
3.85 K. Olszyna, R.D. Cadle, R.G. de Pena: J. Geophys. Res. *84*, 1771-1775 (1979)
3.86 E.E. Ferguson: Geophys. Res. Lett. *5*, 1035-1038 (1978)
3.87 J.J. Huntzicker, R.A. Cary, C-S Ling: Environ. Sci. Technol. *14*, 819-824 (1980)
3.88 R.D. Cadle, R.C. Robbins: Discuss. Faraday Soc. *30*, 155 (1961)
3.89 L.R. Martin, H.S. Judeikis, M. Wun: J. Geophys. Res. *85*, 5511-5518 (1980)
3.90 P. Ausloos, R.E. Rebbert, L. Glasgow: J. Res. Natl. Bur. Stand. *82*, 1-8 (1977)
3.91 D. Keefe, P.J. Noland, T.A. Rich: Proc. R. Ir. Acad. A*60*, 27-45 (1959)
3.92 H. Kojima: Atmos. Environ. *12*, 2363-2368 (1978)
3.93 P. Hamill, R.P. Turco, O.B. Toon, C.S. Kiang, R.C. Whitten: J. Aerosol Sci. (in press, 1982)

Additional References

F. Arnold, R. Fabian, W. Joos: " Measurements of the height variation of sulfuric acid vapor concentrations in the stratosphere". Geophys. Res. Lett. *8*, 293-296 (1981)

A.A. Viggiano, F. Arnold: "Extended sulfuric acid vapor concentration measurements in the stratosphere". Geophys. Res. Lett. *8*, 583-586 (1981)

4. Models of Stratospheric Aerosols and Dust

R. P. Turco

With 16 Figures

This chapter deals with the structure and practical use of stratospheric aerosol models. The general mathematical framework of such models is developed, and several published models are reviewed. Applications and results of previous theoretical aerosol studies are discussed. Model predictions of aerosol physical properties, and of aerosol precursor gas distributions, are compared with measurements. Highlighted are useful insights into the origins and characteristics of aerosols gained through theoretical simulation and analysis. Model predictions of man's impact on the stratospheric aerosol layer, as a result of ongoing aerospace and industrial activities, are also presented.

4.1 Overview

In the short time since the discovery of the stratospheric sulfate layer by JUNGE and co-workers [4.1], an explosive growth has occurred in experimental and theoretical knowledge of the layer. This was due in part to its accessibility from aircraft and balloon platforms, coupled with the feasibility of remote sensing using lidars and satellites, as outlined in Chap. 2. Research was also spurred by the realization that airborne particles can influence weather and climate, atmospheric electricity, air pollution, and the trace composition of the atmosphere, among other things. As a consequence, models are now actively sought to test theories of particle formation and removal, to guide experimentalists in instrument design and measurement strategy, and to utilize the latest theories and observations of aerosol properties to answer questions of practical importance.

The behavior of aerocolloidal systems has long been a subject of interest and curiosity to atmospheric physicists. Thus, in a somewhat disconnected way, theoretical treatments of individual microphysical processes — for example, particle nucleation, coagulation, and sedimentation — have become highly sophisticated. Section 1.3 presents an overview of the basic microphysical processes affecting particles suspended in the atmosphere.

JUNGE et al. [4.1] constructed one of the first stratospheric aerosol models. Although they were aware of the complex origin of the aerosols (oxidation and nucleation of SO_2 gas transported upward from the troposphere), they were restricted to a relatively simple analytical treatment of the problem, which included sedimentation and diffusion of particles, as well as an approximate description of coagulation effects. JUNGE et al. concluded from their studies that these physical processes, plus aerosol growth by condensation, satisfactorily described the evolution of particle distributions in the stratosphere. They were also able to divide the aerosols into three distinct categories according to origin: those with radii <0.1 μm, which appeared to originate in the troposphere, those with radii between 0.1 and 1 μm, which seemed to form within the stratosphere from the oxidation products of SO_2 or H_2S, and those with radii >1μm, which appeared to have an extraterrestrial origin. We now know that, with few exceptions, the JUNGE et al. model qualitatively accounts for the characteristics, sources and sinks, and processes of stratospheric aerosols quite well.

No further model development seems to have been attempted until the mid-1970's. BURGMEIER and BLIFFORD [4.2] constructed a model to simulate aerosol 'aging' due to growth by coagulation and condensation, and removal by sedimentation. They solved a more accurate formulation of the aerosol coagulation equations than JUNGE et al. [4.1], but largely neglected vertical transport. BURGMEIER and BLIFFORD deduced that the stratospheric aerosol size distribution is very sensitive to the particle source function and that the rate of gas-to-particle conversion is $\leq 10^{-20}$ g cm^{-3} s^{-1}.

KRITZ [4.3] undertook a more general analysis of aerosol formation and growth. Vertical transport by sedimentation and diffusion was treated using residence time arguments. KRITZ employed his model to simulate a number of possible scenarios related to the generation of the ambient aerosol layer in order to assess the effects of different physical mechanisms on aerosol concentrations and size distributions. Among other things, he endorsed the idea that the stratospheric aerosols are formed by vapor condensation on Aitken nuclei of tropospheric origin.

ROSEN et al. [4.4] adopted a different approach to modeling in which they used a more realistic formulation of vertical transport processes (i.e., similar to that of [4.1]), but simplified the treatment of particle coagulation and growth mechanisms. In particular, they assumed that the aerosol coagulation kernels were independent of size, and that gas-to-particle conversion occurred in a narrow 'growth' layer at 20 km. Largely as a result of the assumptions made, ROSEN et al. were able to obtain solutions in rough qualitative agreement with many of their earlier observations of particle concentrations and size distributions. They concluded that one-dimensional models are adequate to explain most of the general features of the stratospheric aerosol layer.

UCHINO et al. [4.5] made a thorough analysis of aerosol dynamics in three separate size ranges: A($r \leq 0.01$ μm), B(0.01 μm $\leq r \leq 0.1$ μm), and C($r \geq 0.1$ μm). They found that size A particles are produced mainly by nucleation and coagulation processes, size B particles are most sensitive to tropospheric Aitken nuclei, and size C particles grow primarily by attachment of smaller particles.

TURCO et al. [4.6] constructed the most elaborate model of the globally averaged stratospheric sulfate aerosol layer (also see [4.7]). They incorporated most of the features of the models discussed above, and included some additional features such as a treatment of gas-phase sulfur chemistry and a variable composition for the aerosols. Calculations and results from the TURCO et al. model are highlighted below.

A generalized approach to stratospheric aerosol modeling, appropriate for computer applications, is developed in this chapter. Some of the fine details and subtle interactions of aerosol systems are emphasized. Similar techniques have been developed for studying tropospheric aerosols, and these have been reviewed elsewhere [4.8]. By analyzing a generalized equation which describes aerosol behavior for a wide range of conditions, the approximations implicit in published aerosol calculations can be revealed. The best understanding of a model is achieved through an understanding of the approximations and assumptions used in its construction. WHITTEN et al. [4.9] recently reviewed the structure of stratospheric aerosol models, and this survey draws on and extends that review.

The present chapter is organized as follows. A description of the generalized aerosol continuity equation is given in Sect.4.2. There, approximate solutions of the generalized equation are discussed to illustrate the origin of present-day aerosol models. In Sect.4.3, a modern, comprehensive stratospheric aerosol model is analyzed to emphasize the approaches that may be taken to solve the complex atmospheric aerosol system. Model simulations of stratospheric aerosols are presented to demonstrate the role of aerosols in current scientific research and to illustrate the variety of information that can be gained from theoretical studies. In Sect.4.4, models of other upper-atmospheric particles that may affect stratospheric aerosols, such as meteoric dust, are briefly surveyed. Model predictions of anthropogenic influences on stratospheric aerosols are summarized in Sect.4.5. Concluding remarks are given in Sect.4.6.

4.2 The Generalized Aerosol Continuity Equation

Most stratospheric aerosol models are based on a governing equation which describes particle interactions and dynamics. Such a generalized aerosol continuity equation (GACE), from which most aerosol models may be derived, can be written as

$$\frac{\partial}{\partial t} n(r,\underline{x},t) = S(r,\underline{x},t) - L(r,\underline{x},t)n(r,\underline{x},t)$$
$$- \frac{\partial}{\partial r}[g(r,\underline{x},t)n(r,\underline{x},t)] - \underline{\nabla}_{\underline{x}} \cdot \underline{\phi}_n(r,\underline{x},t)$$

$$- n(r,\underline{x},t) \int_0^\infty K(r,r',\underline{x})n(r',\underline{x},t)dr'$$

$$+ \int_0^{r/2^{1/3}} \left(\frac{r}{r''}\right)^2 K(r',r'',\underline{x})n(r',\underline{x},t)n(r'',\underline{x},t)dr' \qquad (4.1)$$

where $r'' = (r^3 - r'^3)^{1/3}$. In (4.1), n is the particle size distribution; e.g., the number of particles [cm^{-3} μm^{-1}] of radius r[μm] located at spatial position $\underline{x}(x,y,z)$ at time t. K is the particle coagulation, or collection, kernel [cm^3 s^{-1}]. S [particles cm^{-3} μm^{-1} s^{-1}] is the aerosol source function (e.g., due to injection, nucleation, and breakup of particles). L[s^{-1}] is the particle loss rate attributable to washout, rainout, and nucleation processes (note that dust is lost by nucleation, while aerosols are gained). The growth rate of aerosols as a result of vapor condensation is denoted by g[μm s^{-1}], which may be decomposed as follows:

$$g(r,\underline{x},t) = g_0(r,\underline{x})[n_g(\underline{x},t) - n_g^0(r,\underline{x},t)] \qquad (4.2)$$

where g_0 is the growth 'kernel' involving kinematic terms and an 'accommodation' coefficient, n_g is the concentration of the condensing gas [molecules cm^{-3}] in the vicinity of the aerosol, and n_g^0 is the equilibrium vapor concentration of the condensing gas over the aerosol.

The spatial particle flux, ϕ_n[particles cm^{-2} s^{-1} μm^{-1}], is composed of several terms,

$$\underline{\phi}_n(r,\underline{x},t) = \underline{v}(\underline{x},t)n(r,\underline{x},t) - v_s(r,\underline{x}) \hat{z} n(r,\underline{x},t) + \underline{\phi}_D \qquad (4.3)$$

where $\underline{v}$[cm s^{-1}] is the average bulk velocity of the atmosphere, v_s[cm s^{-1}] is the particle sedimentation velocity (due to gravity), $\hat{z}$ is a vertical (upward) unit vector, and $\underline{\phi}_D$ is the diffusive particle flux caused by small-scale turbulent atmospheric motions. The vertical component of the diffusive flux may be put in the form,

$$\phi_{Dz} = \frac{D(z)}{\gamma(z)} \frac{\partial}{\partial z} [\gamma(z)n(r,\underline{x},t)] \qquad (4.4)$$

where D[cm^2 s^{-1}] is the vertical 'eddy' diffusion coefficient and γ is an atmospheric stratification parameter,

$$\gamma(z) = M(z_0)/M(z) \qquad (4.5)$$

where M is the concentration of air molecules [number cm^{-3}].

Note that particle growth by condensation (4.2) involves an interaction between the aerosols and their precursor gases. The precursor gases also control the aerosol source term S through microscopic nucleation processes. For each 'active' precursor gas, therefore, a separate continuity equation is written,

$$\frac{\partial}{\partial t} n_g(\underline{x},t) = P_g(\underline{x},t) - n_g(\underline{x},t)L_g(\underline{x},t)$$
$$+ E_g(\underline{x},t) - C_g(\underline{x},t) - \underline{\nabla}_x \cdot \underline{\phi}_g(\underline{x},t) \quad . \tag{4.6}$$

Here, P_g [molecules cm^{-3} s^{-1}] and L_g [s^{-1}] are the homogeneous chemical production and loss rates, respectively, for the gas (see Chap.3 for a discussion of aerosol precursor gas photochemistry). E_g and C_g represent the particle interaction terms; E_g includes evaporation from particles and surface reactions as sources of molecules, while C_g accounts for vapor condensation and decomposition on surfaces, and vapor incorporation into nucleation germs, as sinks for molecules. ϕ_g is the vapor flux, which has both diffusive and advective components.

If a particular aerosol-gas interaction term — such as the growth rate expressed by (4.2) — depends on a well-defined property of the aerosol — such as its surface area or volume — the related terms in E_g and C_g will contain appropriate integrals over the particle size distribution n. For example, the rate of collection of a gas by particles is, in its simplest form, proportional to the surface area of the particles. In the general case,

$$C_g(\underline{x},t) = n_g(\underline{x},t)\beta_g \int_0^\infty g_0(r,\underline{x})n(r,\underline{x},t)4\pi r^2 dr \quad ,$$
$$E_g(\underline{x},t) = \beta_g \int_0^\infty n_g^0(r,\underline{x},t)g_0(r,\underline{x})n(r,\underline{x},t)4\pi r^2 dr \quad , \tag{4.7}$$

where β_g is a factor which converts the particle growth rate [μm s^{-1}] to a molecular flux.

The atmosphere is a weakly ionized plasma. The particles suspended in the atmosphere will react with the ambient ions and electrons, altering the overall state of electrification. Charge-carrying particles interact differently than neutral particles, depending on the particle size and charge involved. HIRONO et al. [4.10] made simultaneous particle and ion measurements and established a theoretical model to explain the data.

The basic atmospheric plasma equations in the presence of aerosols can be written as follows:

$$\frac{dn_+}{dt} = P - \alpha_e n_e n_+ - \alpha_i n_+ n_- - A_+ n_+ - J_+ n_+$$
$$\frac{dn_e}{dt} = P + \psi + \delta n_- - \beta n_e - \alpha_e n_e n_+ - A_e n_e \tag{4.8}$$
$$\frac{dn_-}{dt} = \beta n_e - \delta n_- - \alpha_i n_- n_+ - A_- n_- - J_- n_- \quad ,$$

where n_+, n_e and n_- are the positive ion, electron, and negative ion concentrations, respectively; P [ion-electron pairs cm^{-3} s^{-1}] is the air ionization rate; α_e and α_i

are the ion-electron and ion-ion recombination coefficients [$cm^3 s^{-1}$], respectively; δ [s^{-1}] is the negative ion detachment coefficient; and β [s^{-1}] is the electron molecular attachment coefficient. The collection rates of ions and electrons by aerosols, $A_{\pm,e}$ [s^{-1}], are functions of particle charge and involve integrals over the particle size distribution. The electron photoemission rate, ψ [electrons cm^{-3} s^{-1}], involves integrals both over wavelength (the integrand being the product of the attenuated solar flux, the particle absorption cross section, and the material electron yield function) and over particle size distribution. The $J_{\pm}$ [s^{-1}] are the positive and negative ion nucleation rates (per ion).

Equations (4.8) are the well-known 'lumped parameter' electrical charge equations. In the real atmosphere, dozens of ion species undergo hundreds of reactions. However, in many applications, (4.8) give an adequate representation of the plasma when the parameters α_e, α_i, δ, and β are properly chosen [4.11]. PARTHASARATHY [4.12] and TURCO et al. [4.13] describe the solutions of (4.8) in the presence of a highly dispersed aerosol; the detailed expressions are not reproduced here. In the stratosphere, the electron component of the plasma is small, and can generally be neglected.

Although the system of equations (4.1-8) is quite complex and requires extensive simplification to obtain tractable solutions, the expressions already contain a number of important approximations. For example, only one type of particle is dealt with in these equations, and variations in particle composition are neglected. A system of several different types of particles would require additional GACEs, including appropriate interaction terms. Moreover, the composition of the particles, which can affect the microphysical properties of the aerosols, must be taken into account. A full treatment of aerosol composition variations can quickly lead to an unmanageable increase in the dimensionality of the particle size distribution, n. Nonetheless, (4.1) can be extended to accommodate the additional degrees of freedom associated with particle type and composition.

Other, more subtle assumptions implied in (4.1-8) are that g_0, K, and v_s are independent of time, and that slip-free advection of small particles holds. Also, the (occasionally considerable) microphysical effects of latent heat, phoretic forces, and electrical charge are explicitly ignored although the last effects are partially described by (4.8).

Uncertainties in aerosol models also result from the difficultes encountered in defining physical 'constants' such as coagulation kernels, vapor pressures, and nucleation rates. Discussions of these problems are presented in Chap.1 and Chap.3, and the references cited therein.

In spite of such approximations and uncertainties, (4.1-8) represent a useful formal approach to the analysis of aerosol physics problems. Unfortunately, these equations have never been satisfactorily incorporated into aerosol models with two or three spatial dimensions because of the inordinate demand on computer time and storage. Thus, most of the models available for the present discussion are one-dimensional (vertical column) models, or even simpler 'box' models.

Two decades ago, fast computers were not readily available and analytical solutions of the aerosol continuity equation were sought. One can readily envision the simplified forms of (4.1) which are amenable to the analytical approach. Generally, these forms involve reduction to low spatial dimensionality; inclusion of only one or two physical processes; assumption of the steady state; adoption of simplified or idealized source, sink, and interaction terms; omission of transport; and insertion of fixed particle sizes or size distributions. Such models, which are described in several excellent monographs on aerosol systems [4.8,14-17], are not pursued in detail here. A particularly simple and interesting example of an analytical solution is provided by the steady-state equation of a vertical column of noninteracting monodispersed aerosols undergoing diffusion and sedimentation. The net vertical particle flux in this case is

$$\phi_{Ds} = \phi_D + \phi_s = -\frac{D}{\gamma\mu} \frac{\partial}{\partial z}(\gamma\mu n) \quad , \tag{4.9}$$

where μ is given by

$$\mu = \exp\left[\int_0^z \frac{v_s(z')}{D(z')} dz'\right] \quad . \tag{4.10}$$

Obviously, in equilibrium, with zero net flux, n is distributed with height as $(\gamma\mu)^{-1}$. Expression (4.9) is useful in formulating aerosol models because it can be used to avoid errors caused by numerical 'diffusion' (see below).

Models of gas and particle dynamics designed for computers may be cast in one of two general numerical forms. Eulerian models utilize fixed spatial and particle-radius grids, and finite differencing is performed over these grids. Lagrangian models, on the other hand, follow specific particle streamlines in space and size coordinates as they evolve with time. Only Eulerian models are discussed here, as no practical Lagrangian models for stratospheric aerosols have yet been devised.

The aerosol continuity equation (4.1) may, alternatively, be analyzed using the method of moments. In this approach, an operator, $M_i = \int_0^\infty dr\, f_i(r)$ ($i = 0,1,\ldots,I$), is applied to (4.1), resulting in a set of I differential equations for the moments, $M(f_i)$, of the size distribution n; f_i represents an appropriate set of basis functions, for example, r^{i-1}. The moments, although independent of radius, basically describe the aerosol size dispersion. This technique is a powerful tool for studying aerosol systems which exhibit size distributions with simple modal structures. The principles of the application of moment theory to aerosol problems are discussed elsewhere [4.8,16]. The use of moments to characterize the size distribution of the trace components found in aerosols is mentioned later.

To proceed with computer analysis, the GACE — e.g., (4.1) — must be recast as a finite-difference equation. The standard techniques for making the transformation from a continuous to a discrete formulation are documented in numerous texts on numerical methods. Generally, in constructing discrete equations, one must strike a careful balance between stable implicit forms and conservative explicit forms. The choice depends to a large extent on the problem to be solved.

Numerical diffusion in the finite-difference solutions of particle advection, sedimentation, and growth terms poses a serious difficulty in finding the solutions of GACEs. Various researchers, whose work is described below, have developed schemes to overcome errors due to numerical diffusion. The accurate and efficient solution of the coagulation integrals over a wide particle size dispersion is another source of difficulty in numerical aerosol modeling. Problems such as these are addressed in the modeling studies outlined in the next section.

4.3 Aerosol Models

A host of stratospheric aerosol models have been developed in recent years. Some of these models are reviewed in Sect.4.1. In almost every case, the model is either a 'box' model or a one-dimensional model. Two-dimensional models utilizing realistic particle microphysics are not yet available; nevertheless, several pioneering two-dimensional studies of the dispersion of volcanic eruption clouds have been made by CADLE and associates [4.18,19]. The one-dimensional model of TURCO et al. [4.6] and TOON et al. [4.7] is the most comprehensive aerosol simulation published to date, combining and refining elements from previous aerosol models. A detailed description of the aerosol physics and chemistry employed, and of the numerical analogs of the aerosol equations solved, can be found in [4.20,21]. Because this model provides a focus for current aerosol simulation efforts, most of the subsequent discussion concentrates on its characteristics and predictive capabilities.

The one-dimensional model of TURCO et al. [4.6] incorporates all of the physical elements and interactions expressed in (4.1-8). These elements and interactions are summarized schematically in Fig.4.1. The particle size range modeled is variable, but generally extends from molecular sizes (~3 Å radius) to several microns. To cover this broad size span economically, particles are placed in geometrically increasing size bins. The particle volume doubles from one size bin to the next, i.e., the radius increases by a factor of $2^{1/3}$. TURCO et al. [4.6,20] and TOON et al. [4.7,21] showed that particle microphysical interactions can be accurately simulated with such a size grid. The nominal height range and vertical resolution of the Turco et al. model are 0-60 km and 2 km, respectively. The background air temperatures and densities are generally assigned fixed values corresponding to a standard model atmosphere.

The aerosol source function $S(r,z,t)$ in the Turco et al. model includes terms due to homogeneous and ion nucleation of H_2SO_4/H_2O binary solutions [4.22], and terms due to heterogeneous nucleation of sulfuric acid and water on a variety of

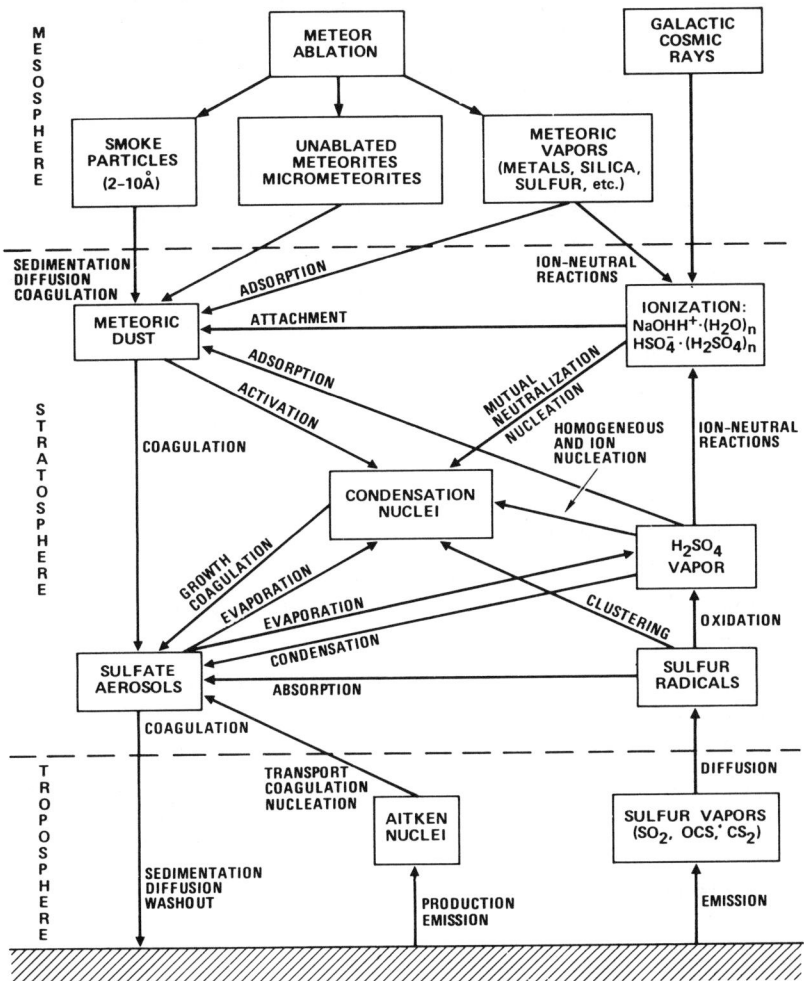

Fig.4.1. A schematic outline of the physical and chemical processes which are included in the stratospheric aerosol model of TURCO et al. [4.6]. Meteoric debris, positive and negative ions, gas-phase sulfur radicals, and tropospheric Aitken particles all serve as nucleation sources for the aerosols, which grow, evaporate, coagulate, and fall and diffuse in the vertical direction. [4.6]

solid particle surfaces including preexisting Aitken nuclei [4.23,24], meteoric dust [4.25,26], and soot from aircraft engines and aluminum oxide particles from rocket motors [4.27]. All of the aerosol nucleation rates are calculated using the 'classical' bulk thermodynamic representation of small sulfuric acid droplets [4.28]. The nucleation rates are also calculated interactively with the other particle processes in the model. It was found that such interactions are crucial to an overall description of stratospheric aerosol nucleation [4.28]. For example, ion nucleation consumes free ions, which are limited in number by the galactic cosmic ray ionization source P in (4.8). In addition, each nucleation consumes H_2SO_4 vapor, as determined by the rate of activation of critical nuclei and the composition of the nuclei; the associated acid vapor loss is accounted for in the term C_g in (4.6). (Water vapor, on the other hand, is so abundant relative to sulfur in the stratosphere that its concentration is not affected by aerosol formation).

The concentrations of solid particles, or condensation nuclei (cn), are analyzed using a separate GACE, in view of their unique microphysical properties (sources, composition, size distribution, and nucleation rate). Although TURCO et al. [4.6] lumped together cn of all types, two or more components (e.g., tropospheric cn and meteoric dust) could be traced individually [4.26]. The evolution of the cn size distribution is affected by incorporation into aerosols as 'cores' as a result of nucleation and coagulation, and by release of the cores upon evaporation of droplets. The coalescence of the absorbed cn (cores) within the droplets affects their overall size spectrum. TURCO et al. [4.6] accounted for core coalescence by solving an additional pair of equations — derived from (4.1) — that apply to the droplet core *first* and *second* volume moments; this technique was shown to be quite precise [4.20].

The aerosol loss term $L(r,z)$ in the Turco et al. model includes the effects of tropospheric washout and rainout processes, based on simple parameterizations of the observed average rates of removal [4.7]. The growth kernel $g_0(r,z)$ and coagulation kernel $K(r,r',z)$ are calculated using the expressions derived by HAMILL et al. [4.23] and HOPPEL [4.29] for the former, and by FUCHS [4.14,15] for the latter.

According to the previous discussion, the proper conservation of mass in a model requires an accurate balance between the aerosol cores and cn, and between the budgets of condensed and vaporous sulfur compounds. The fully interactive model of TURCO et al. [4.6,7] achieves this goal.

4.3.1 The Simulated Distributions of Aerosol Precursor Gases

Stratospheric aerosols are formed as a by-product of the decomposition of sulfur-bearing gases at high altitudes. The precursor gases — principally SO_2, OCS, and CS_2 — originate mainly in the troposphere. Normally, these gases are slowly transported into the stratosphere by diffusion and large-scale motions. However, volcanoes periodically inject immense quantities of sulfur-bearing compounds above the tropopause in a matter of only hours or days. The chemistry linking the injected sulfur gases with particle formation is reviewed in Chap.3.

TURCO et al. [4.30,31] made a detailed study of the tropospheric OCS cycle and the stratospheric sulfur balance during volcanically quiescent times. Figure 4.2 illustrates the estimated flow of sulfur into and out of the stratosphere. The CS_2 flux may be greatly overestimated in Fig.4.2 because recent CS_2 measurements indicate very low background concentrations above the boundary layer [4.32]. Thus, OCS

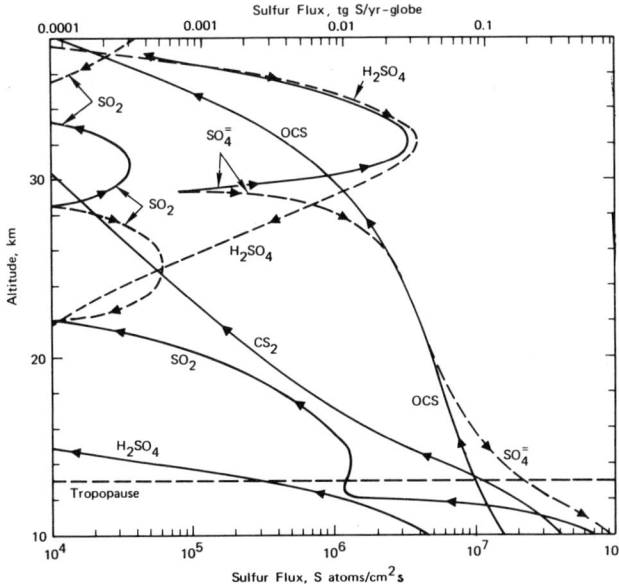

Fig.4.2. The balance of stratospheric sulfur gases and sulfate aerosols in a one-dimensional model. Equivalent sulfur-atom fluxes are given. The arrows indicate direction of sulfur flow. [4.30]

appears to be the dominant sulfur source for the stratospheric aerosol layer during periods of low volcanic activity, as originally proposed by CRUTZEN [4.33]. Sulfuric acid aerosols which diffuse above ~30 km are thermodynamically unstable, and evaporate. This explains the downward flux of H_2SO_4 seen at these altitudes in Fig.4.2. Simulations of the effects of volcanoes on stratospheric aerosols are discussed in Sect.4.3.3.

The concentration profiles of sulfur gases in the tropsphere and stratosphere are illustrated in Fig.4.3. Notice that OCS is the predominant sulfur-bearing constituent in the atmosphere of the Earth. The concentration curves marked 'R' in Fig.4.3 roughly correspond to the flux curves given in Fig.4.2. TURCO et al. [4.43] used a combination of SO_2, OCS, and sulfate aerosol data to suggest that stratospheric OH concentrations are lower than photochemical ozone models were predicting (in 1980). Such work demonstrates the close coupling between sulfur vapor abundances, stratospheric photochemical cycles, and sulfur particulate properties, and emphasizes the important role of aerosols in stratospheric science.

Sulfuric acid vapor concentration is difficult to calculate because it is controlled by heterogeneous processes. Figure 4.4 shows several profiles calculated with an interactive H_2SO_4-solution/H_2SO_4-vapor model [4.26]. Although reliable H_2SO_4 measurements have not yet been made, the preliminary measurement at 37 km shown in Fig.4.4 suggests that H_2SO_4 concentrations may be greatly overestimated by models in which the H_2SO_4 sink on aerosols is reversible, i.e., in which the aerosols can evaporate at high altitudes. On the other hand, new observations of the H_2SO_4 vapor profile between 23 and 33 km are in excellent accord with the profiles of Fig.4.4 [4.45]. TURCO et al. [4.26] suggested that meteoric metals may

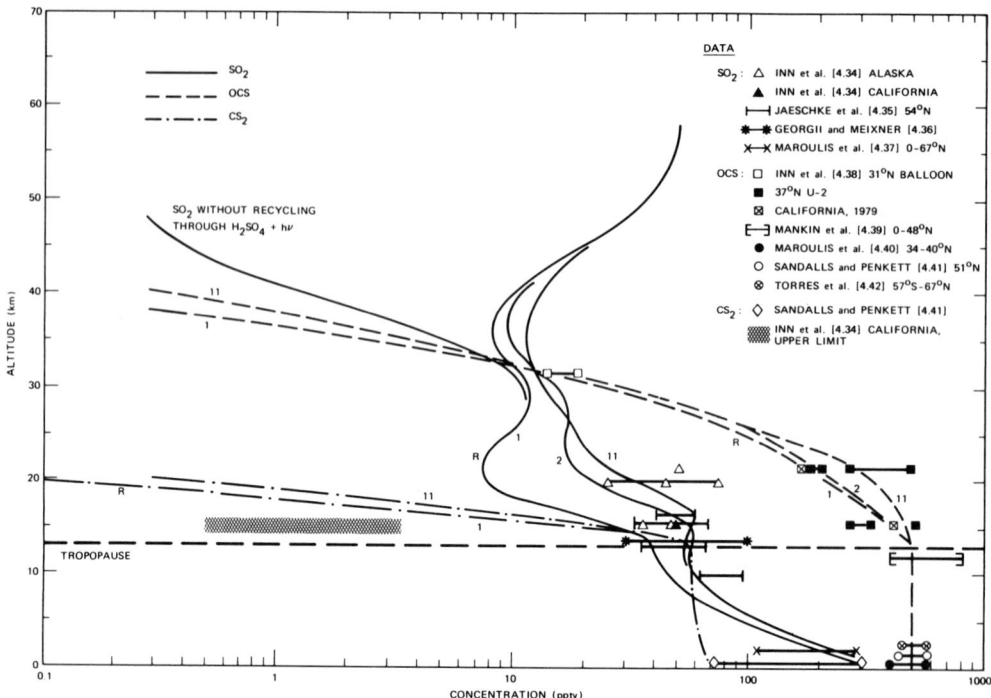

Fig.4.3. Calculated and observed sulfur gas concentrations in the stratosphere. Profiles numbered '1' refer to a high concentration of OH radicals, those numbered '11', to a low concentration of OH. [4.43]

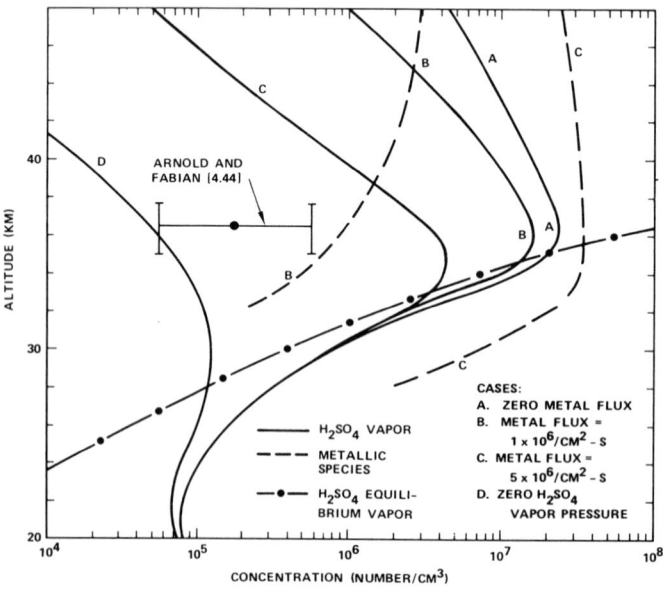

Fig.4.4. Predicted stratospheric distributions of H_2SO_4 vapor and metallic meteoric vapors [4.26]

neutralize H_2SO_4 above 30 km. Indeed, the calculations of meteoric metal and sulfuric acid vapor profiles given in Fig.4.4 indicate that heterogeneous neutralization processes may be important at the uppermost extent of the aerosol layer, a fact which could reconcile the two sets of H_2SO_4 measurements [4.44,45].

4.3.2 Aerosol Nucleation Models

Over the years, theoretical treatments of stratospheric aerosol nucleation processes have grown increasingly sophisticated. Basic information has been forthcoming from laboratory studies of the properties of charged and neutral small molecular clusters [4.46,47]. HAMILL et al. [4.22] applied the principles of classical nucleation theory to calculate H_2SO_4-H_2O heteromolecular homogeneous nucleation rates under stratospheric conditions. HIDY et al. [4.48] recently reevaluated homogeneous aerosol nucleation rates in the lower stratosphere. CHAN and MOHNEN [4.49] developed a model for aerosol nucleation onto ions, and argued that aerosol formation by this mechanism is probably negligible. FRIEND et al. [4.50] estimated particle production rates due to the self-agglomeration of sulfur radicals (H_2SO_4 precursor molecules). ARNOLD [4.51] measured stratospheric ion composition and concluded that ion-ion recombination clusters and multiion complexes might contribute to aerosol formation.

HAMILL et al. [4.28] combined all of these nucleation processes in a single model, and added possible aerosol sources due to the heterogeneous nucleation of Aitken particles and meteoric debris. Each nucleation process, moreover, was coupled interactively to the aerosol physics and chemistry. Typical predictions made with the interactive model are shown in Fig.4.5. The competition between different nucleation mechanisms and the presence of preexisting aerosols were found to have significant effects on the predicted nucleation rates. For example, the surface area provided by the ambient stratospheric aerosols restricted the H_2SO_4-H_2O supersaturations to such low values that homogeneous aerosol nucleation could not proceed in the stratosphere. The most favorable region for homogeneous nucleation occurred in the upper troposphere (∼10 km), where temperatures are low and humidities are often high. HAMILL et al. [4.28] did not take into account the possible kinetic limitation to homogeneous H_2SO_4-H_2O nucleation imposed by the finite lifetime of growing subcritical acid-water clusters against coagulation with preexisting aerosols. Near 10 km, the predicted sulfuric acid vapor concentration is low enough ($\sim 10^6$ cm^{-3}) that homogeneous nucleation may be kinetically limited.

Ion nucleation also appears to be kinetically limited because of the finite lifetime of ions against neutralization. Here, HAMILL et al. [4.28] assumed that, if an ion cluster of subcritical size is neutralized by another ion of opposite charge, the resulting neutral cluster is unstable and evaporates. HAMILL et al., allowing for the kinetic limitation to ion nucleation in their calculations, found that atmospheric ion nucleation is inhibited principally by the low concentrations

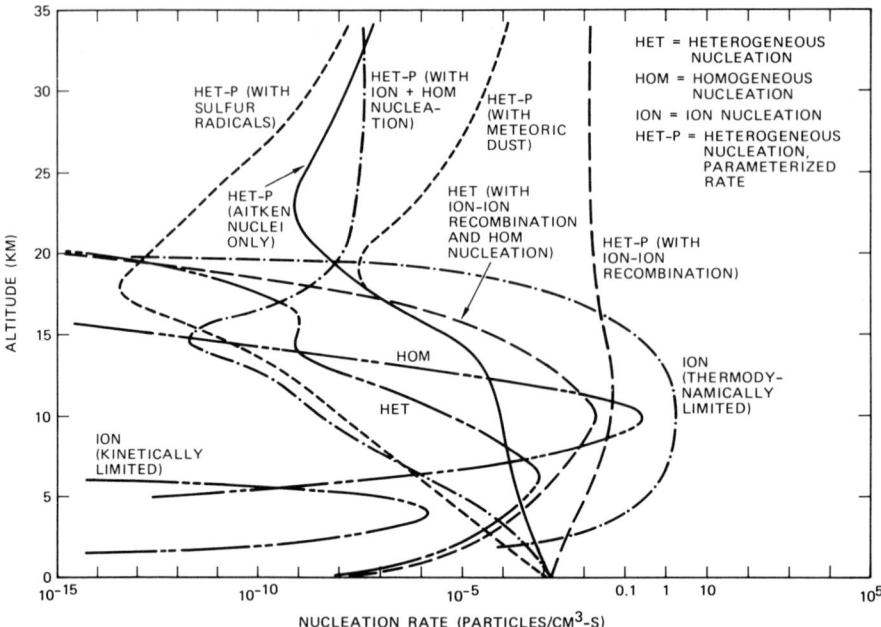

Fig.4.5. Heteromolecular H_2SO_4-H_2O nucleation rates. 'Parameterized' heterogeneous nucleation assumes a nucleation time of 10^6 s for any nucleus in a supersaturated environment

of H_2SO_4 molecules in the presence of the background aerosols (note, however, that the classical thermodynamic equilibrium nucleation theory does not strictly apply in such situations). In Fig.4.5, ion nucleation is seen to be negligible almost everywhere in the ambient atmosphere when kinetic factors are taken into account.

The model adopted by HAMILL et al. [4.28] also suggests that heterogeneous H_2SO_4-H_2O nucleation is ineffective in the stratosphere (Fig.4.5). Nucleation on dust surfaces, however, is very sensitive to a variety of uncertain parameters, including the 'contact' angle. Thus, heterogeneous nucleation may occur more freely in the stratosphere than the theoretical results of HAMILL et al. indicate. For example, the nucleation curves in Fig.4.5 which correspond to *parameterized* (fast heterogeneous) nucleation demonstrate the *potential* importance of this process. Otherwise, it may be inferred from Fig.4.5 that most new stratospheric aerosols originate in the upper troposphere, and are injected across the tropopause.

4.3.3 Calculated Properties of the Aerosols

The measured properties of stratospheric aerosols obtained from a variety of field instruments (Chap.2) provide strict constraints on model predictions. TOON et al. [4.7] made extensive comparisons between model calculations and in situ observations for a variety of parameters, including the total particle mixing ratio, the sulfate mass mixing ratio, the large particle (r > 0.15 μm) mixing ratio, the concentration

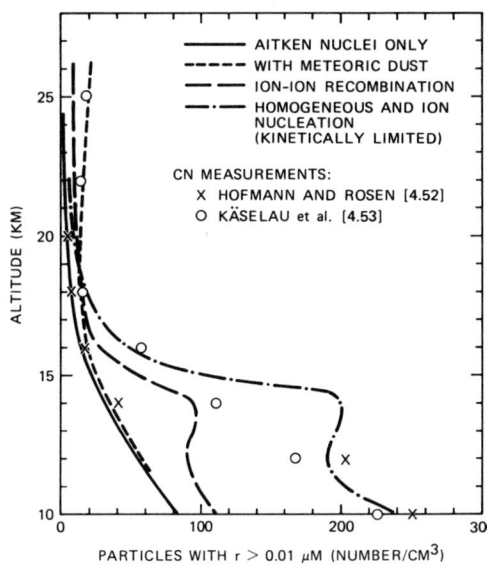

Fig.4.6. Total aerosol particle concentrations in the lower stratosphere (i.e., particles with radii >0.01 μm). Model predictions (lines) are compared with data

ratio of the number >0.15 μm / number >0.25 μm, and the particle composition and size distribution. They found generally good agreement in each case.

Measurements of the total particle concentration obtained with present-day condensation nuclei counters are probably limited to particle sizes $\gtrsim 0.01$ μm radius. In Fig.4.6, typical observational values are compared with the model predictions of TURCO et al. [4.54] and HAMILL et al. [4.28]. The comparison suggests that many of the aerosols detected in the upper troposphere may be generated locally, for example, by homogeneous nucleation. An Aitken nuclei source at the ground also supplies particles to the stratosphere. However, these cn are effectively removed by nucleation and rainout below 10 km.

Figure 4.7 contrasts calculated and observed sulfate mass mixing ratios. Considering the variability of the measurements, the model calculations provide a reasonable fit to the data. TURCO et al. [4.26] predicted that the continual influx of extraterrestrial (meteoric) debris has a small, but perceptible, effect on the aerosol mass (Fig.4.7). Note that, at high altitudes (above ~32 km), sulfuric acid vapor contributes significantly to the total sulfate mass; above 35 km, H_2SO_4 vapor is the dominant sulfate compound (notwithstanding possible absorption and neutralization of H_2SO_4 on meteoritic particles [4.26]).

Calculated and observed aerosol particle size ratios are illustrated in Fig.4.8. The ratio typically has values ~4-6 below 25 km. Above 25 km, the behavior of the size ratio is uncertain. The measurements of HOFMANN and ROSEN [4.58] suggest a rapidly increasing ratio with height, but the scatter and uncertainty in the data at these altitudes preclude a firm conclusion. The theoretical simulations of TURCO et al. [4.26] show a sharp increase in the size ratio above 35 km in a model assuming *pure* sulfuric acid droplets, because the droplets evaporate rapidly above

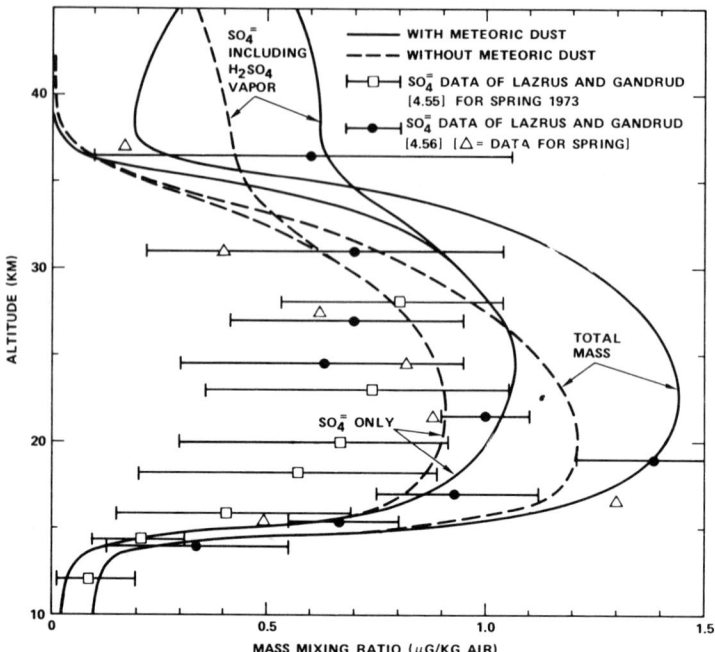

Fig.4.7. Particulate mass mixing ratios calculated with and without meteoric dust. The corresponding $SO_4^=$ mass mixing ratios are also shown, and are compared to the data in [4.55,56]. The measurements in [4.55] were made in spring 1973 in the northern hemisphere. Data points correspond roughly to average values at a fixed elevation with respect to the mean tropopause level. The range of individual measurements is indicated by a crossbar. The $SO_4^=$ data points in [4.56] represent average values of measurements made during 1976, with one standard deviation indicated by a crossbar. Average spring 1976 data, more appropriate for comparison with the model, are plotted as well (standard deviations for these points were not published). The 1976 measurements may have been influenced to some extent by the eruption of Volcan Fuego (14.5°N) in October 1974. [4.26]

35 km. This point of view (i.e., volatile aerosols up to 35 km) is supported by recent measurements of the H_2SO_4 vapor profile [4.45]. On the other hand, TURCO et al. [4.26] noted that, if the aerosols are nonvolatile, a steady increase in the size ratio begins near 25 km (where sedimentation overwhelms diffusion), which is in close agreement with the data of HOFMANN and ROSEN (see Fig.4.8).

Measurements suggest that the composition of the stratospheric aerosols is basically 75% sulfuric acid aqueous solution [4.59] with an admixture of solid granules and dissolved sulfates [4.60,61]. Some of the material in the aerosols may have an extraterrestrial origin. Figure 4.9 illustrates the possible meteoric composition of stratospheric particles of different sizes at several altitudes [4.26]. These calculations agree with observational data in that meteoric debris dominates the particles above ~1-μm radius. The predictions also suggest that meteoric "smoke" (tiny particles recondensed from ablated meteor vapors) dominates the smallest aerosols ($\lesssim$ 0.01 μm), notwithstanding other sources of particles in this size range

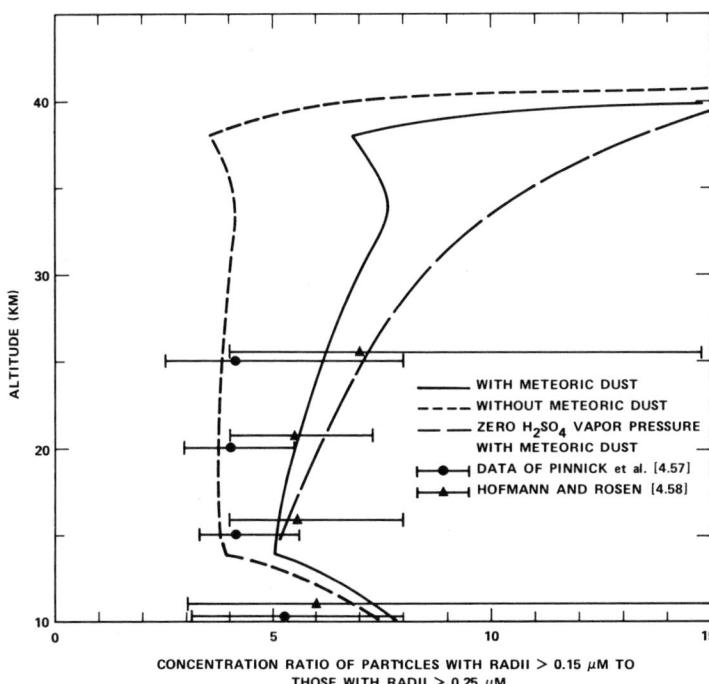

Fig.4.8. Particle size ratios calculated with and without meteoric dust. For comparison, measurements from [4.57] made worldwide between 1971 and 1974, and [4.58] taken at Laramie, WY, during 1978 and 1979, are shown. In each case, an average value and the range of individual measurements are given. The observations from [4.58] are most appropriate for the background aerosol layer in the absence of volcanic activity. [4.26]

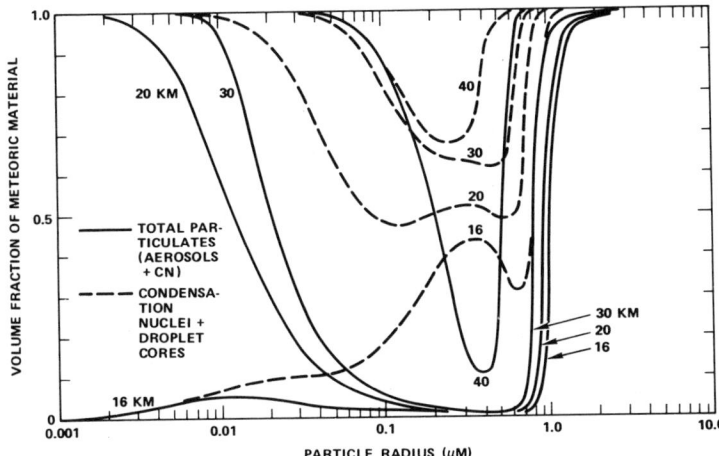

Fig.4.9. Fractional volume of meteoric material in condensation nuclei *plus* aerosol droplets, and in condensation nuclei *plus* aerosol droplet 'cores', as a function of particle size at several altitudes. [4.26]

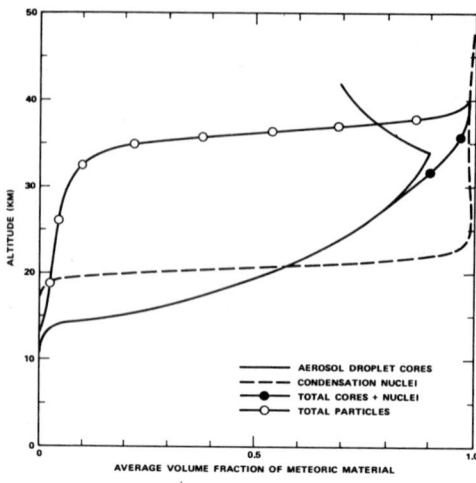

Fig.4.10. Average meteoric composition of stratospheric particles as a function of altitude [4.26]

(e.g., sulfur radical clusters, ion-ion recombination products and Aitken nuclei).

The predicted fractional meteoric composition of different categories of stratospheric particles (aerosols, cn, droplet cores, and total particles) is shown in Fig.4.10 as a function of altitude. Note that, while the cn and cores may be dominated by meteoric dust above 20 km, the total particulate has a very low meteoric content everywhere below 35 km. Accordingly, the detection of meteoric debris in fine stratospheric aerosols may require unusually sensitive techniques. Such results clearly demonstrate that multicomponent aerosol models, which predict the detailed composition of aerosols, can be usefully applied to the design of stratospheric experiments.

Models of ambient stratospheric aerosols have recently been extended to study the properties of volcanic eruption clouds. TURCO et al. [4.62] constructed a model to describe the stratospheric clouds produced by the Mt. St. Helens eruptions of May and June 1980. They used in situ measurements of SO_2, SO_4^{2-}, OCS, H_2O and ash to constrain the model calculations, and showed that, by accounting for the expansion and chemistry of the clouds, most of the observational data could be qualitatively explained (Chap.5 discusses the radiative properties of volcanic clouds).

Figure 4.11 illustrates the first detailed volcanic aerosol size distributions calculated for the Mt. St. Helens eruption of 18 May, 1980. TURCO et al. [4.62] predicted a multimodal aerosol size dispersion. A nucleation mode (not shown in Fig.4.11) dominated the size range below ∼0.01 μm. Ambient aerosols that were mixed into the expanding cloud provided the source of the aerosols between ∼0.01 and 0.1 μm; these aerosols grew substantially by sulfur vapor condensation within the cloud. Two size modes of volcanic ash were also identified. Fine ash particles were distributed between ∼0.1 and 3 μm [4.63], and large ash particles between ∼3 and 30 μm [4.64]. The large ash particles were removed rather quickly by sedimen-

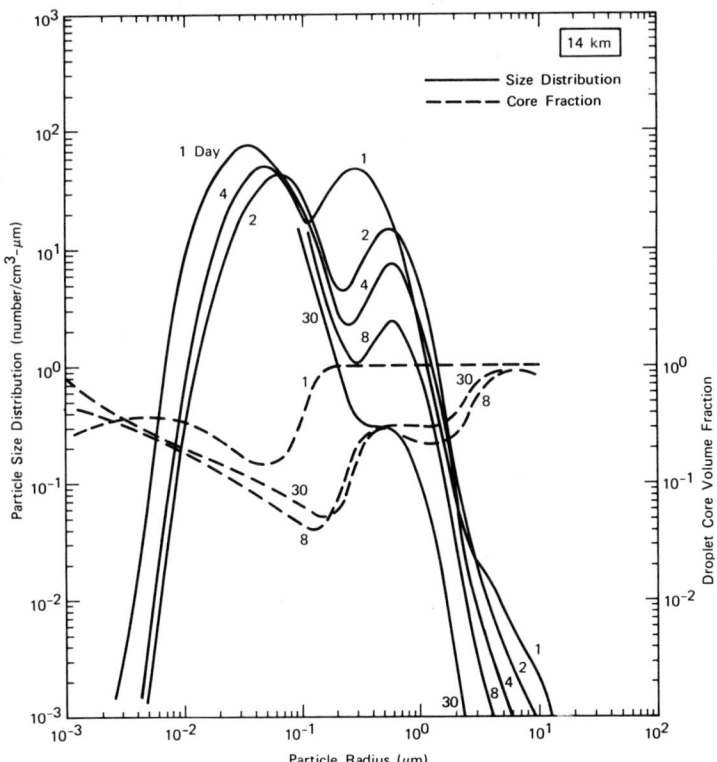

Fig.4.11. Predicted size distributions of volcanic cloud aerosols at 14 km after the 18 May 1980 Mt. St. Helens eruption. The size distributions after 1, 2, 4, 8, and 30 days are shown. Also illustrated are the aerosol droplet 'core' volume fractions after 1, 8, and 30 days. The calculations include heterogeneous sulfur chemistry, but not homogeneous sulfuric acid nucleation. [4.62]

tation (within a few days). The smaller ash grains remained suspended in the stratosphere for several weeks. During this time, the ash catalyzed a SO_2-to-SO_4^{2-} conversion, and grew in size accordingly. In Fig.4.11, the size mode originally composed of fine ash is seen to be ~70% sulfuric acid solution after ~1 week. Aerosol size modes similar to those shown in Fig.4.11 were observed by FARLOW et al. [4.63], CHUAN et al. [4.65], and HOFMANN and ROSEN [4.66]. Despite these encouraging initial results, the physicochemical theory of volcanic aerosol formation will require extensive investigation in the future.

4.4 Models of Upper Atmospheric Dust

While the subject of the present chapter is the stratospheric aerosol layer, it is worthwhile to consider briefly models of the other particulates which reside in and above the stratosphere, and which may affect aerosol composition and properties. The generalized aerosol continuity equation (GACE), of course, applies to these particles as well as to sulfate aerosols. On a grander scale, one might envision a unifying theory that physically couples all of the particulates found in the upper atmosphere; the theory would treat sulfate aerosols, meteoric dust, nacreous clouds, and noctilucent clouds. Moreover, by logical extension, a unifying theory could be developed to explain the high-altitude clouds and hazes on other planets and moons.

Meteoric debris is a prominent component of stratospheric aerosols, and may contribute both to the background cn level and the trace composition of the aerosols (Sects.4.3.2 and 4.3.3). HUNTEN et al. [4.25] made the first detailed microphysical analysis of the meteoric dust size and height distributions in the upper atmosphere. Their model is based on the GACE of Sect.4.2. HUNTEN et al. [4.25] assumed that meteoric 'smoke' is generated in meteor trails. They considered the entire size spectrum of the incoming meteoric debris, including micrometeorites. Typical calculated meteoric dust size distributions are shown in Fig.4.12. Below ∼0.1-μm radius, the size distribution is dominated by coagulated smoke particles. Above ∼0.1 μm, micrometeorites (which do not ablate completely upon entering the atmosphere) determine the size distribution. The predicted particle sizes fit the current meager data base on meteoric dust [4.67].

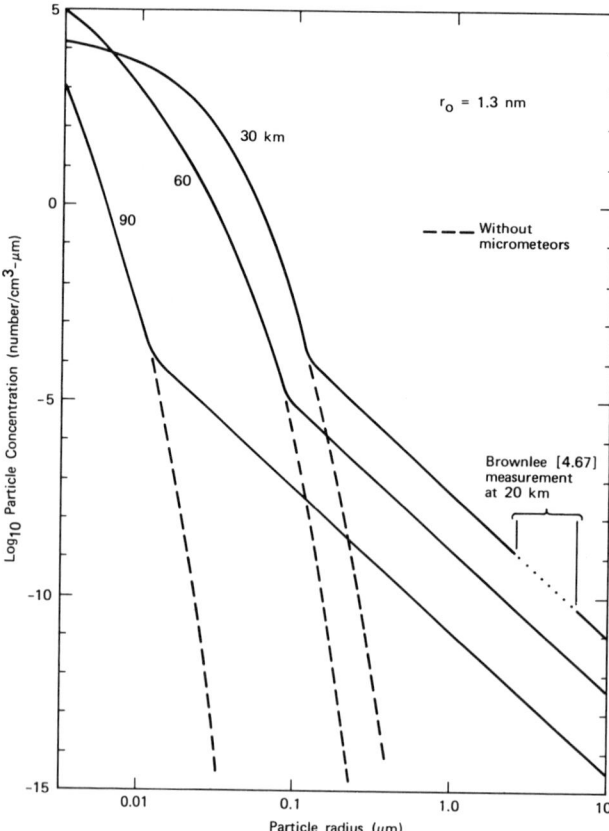

Fig.4.12. Calculated meteoric dust size distributions at several altitudes for an initial meteoric smoke size r_0 of 1.3 nm. Also shown are the size distributions of micrometeorites. [4.25]

Meteoric dust may also be involved in the formation of noctilucent clouds (NLC). The observational properties of NLCs have been reviewed by FOGLE and HAURWITZ [4.68], TOON and FARLOW [4.69], and AVASTE et al. [4.70]. Attempts to model these clouds are described in papers by WITT [4.71], REID [4.72], and TURCO et al. [4.13].

It is noteworthy that the general principles of aerosol dynamics apply to many of the clouds and aerosols detected on the other worlds of our solar system. For example, TOON et al. [4.73] constructed a model of the Titan clouds using a GACE (Eq.4.1), and were able to predict in detail the properties of the clouds and make comprehensive evaluations of the available data.

4.5 Anthropogenic Perturbations of the Aerosols

One of the most useful applications of aerosol models, after proper validation procedures have been carried out, is the assessment of the role which aerosols play in air pollution. Stratospheric aerosols, if enhanced by man's activities, can affect the climate of Earth by altering the transfer of radiation through the atmosphere (Chap.5). Recently, calculations were carried out to estimate the effects that supersonic transport (SST) SO_2 and soot emissions [4.27,74,75], space shuttle Al_2O_3 particulate emissions [4.27], and industrial OCS emissions [4.30,31] have on stratospheric aerosols. Of course, aerosol models based on a GACE can be used to study the complex height-, size-, and time-dependent pollution scenarios related to these sources.

Figure 4.13 shows the predicted effects on aerosols of soot and SO_2 emissions by a fleet of 300 advanced SST's. Although it is not obvious in the figure, the mass and optical depth of the aerosol layer are increased by ~10%-20% by the SST's. Soot contributes mainly to the smallest aerosol sizes (<0.01 μm). The injected SO_2, upon chemical transformation into H_2SO_4 (or other nonvolatile sulfur compounds), condenses on preexisting aerosols, which consequently expand in size. Figure 4.14 shows the corresponding change in the large aerosol particle (r > 0.15 μm) mixing ratio caused by the SST's (the large particles also happen to be the optically active aerosols). An increase of about 10%-20% is apparent for heavy SST traffic. Comprehensive aerosol models yield detailed particle size, height, and composition distributions with which accurate radiation transport calculations can be made to forecast climate variations. In the case of the SST perturbations just described, a global surface cooling of less than 0.01 K was projected [4.27].

HOFMANN et al. [4.74] suggested that the space shuttle effluence of numerous small aluminum oxide particles might eventually generate many additional large particles in the stratosphere. Indeed the calculations in Fig.4.15 indicate that a substantial increase in the total number of particles can occur. However, TURCO et al. [4.27], using a detailed aerosol model, demonstrated that only a very small

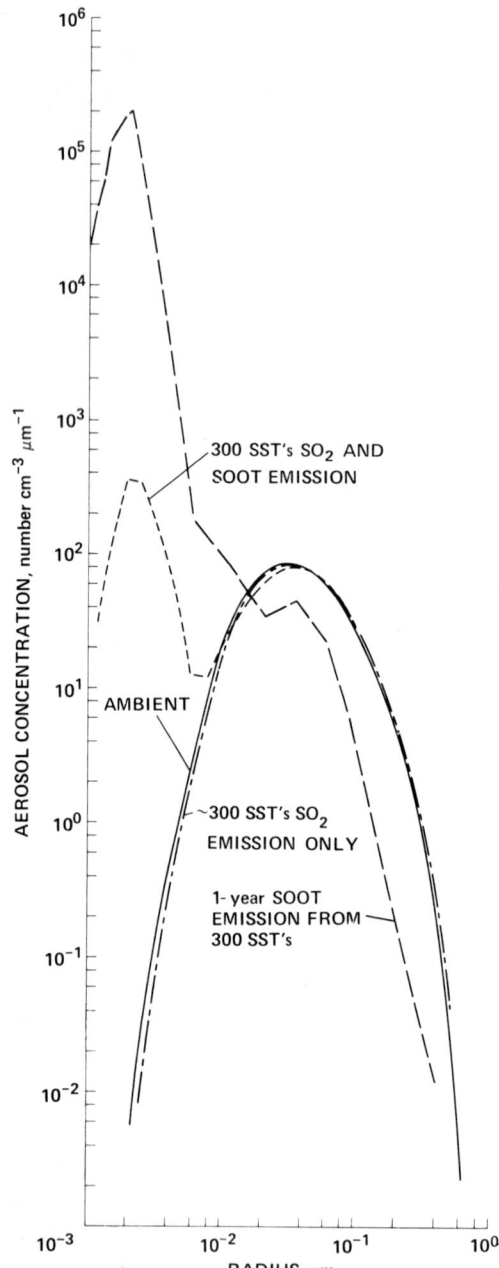

Fig.4.13. Predicted aerosol particle size distributions at 20 km for SST exhaust injection of SO_2 and soot. The global injection rates are: SO_x (2.9×10^7 kg SO_2/yr); soot (8.7×10^6 kg soot/yr). Also shown is the quantity of soot released at 20 km during one year by the (assumed) SST fleet (this curve also defines the size distribution of the injected soot). [4.27]

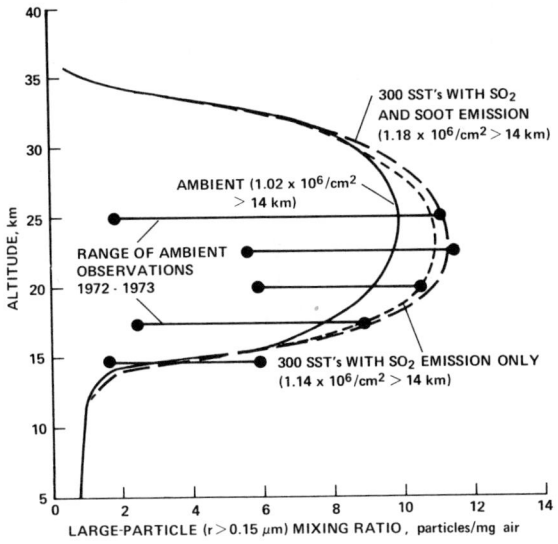

Fig.4.14. SST exhaust effects on large aerosol particles. Shown are calculated steady-state large particle ($r > 0.15$ μm) mixing ratios in the stratosphere for 20-km SST flight, both with and without soot emission. The ambient model large particle mixing ratio profile and some observational data [4.76,77] are shown for comparison. For each calculated profile, the total stratospheric column concentration of large particles is indicated in brackets [4.27]

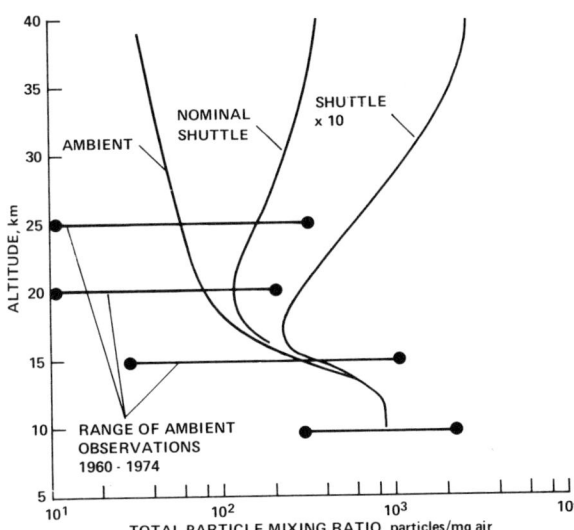

Fig.4.15. Space shuttle effects on the mixing ratio of total stratospheric particles. The ambient model prediction and the range of observational data [4.1,53] are also indicated [4.27]

fraction of the injected particles could grow to large size. The population of large particles in the layer, in fact, was found to be governed by the mass input to the layer to a much greater degree than by the number of injected seed particles.

TURCO et al. [4.30,31] considered in detail the atmospheric budgets of OCS and CS_2. Both gases can contribute to the formation of stratospheric aerosols [4.31,33]. Man's emissions of OCS, associated with fossil fuel refining and combustion, appear to represent a substantial portion of the global OCS budget. The atmospheric lifetime of OCS is undetermined, but is probably $\gtrsim 1$ year. Therefore, increasing consumption of coal and oil worldwide can be expected to increase the atmospheric

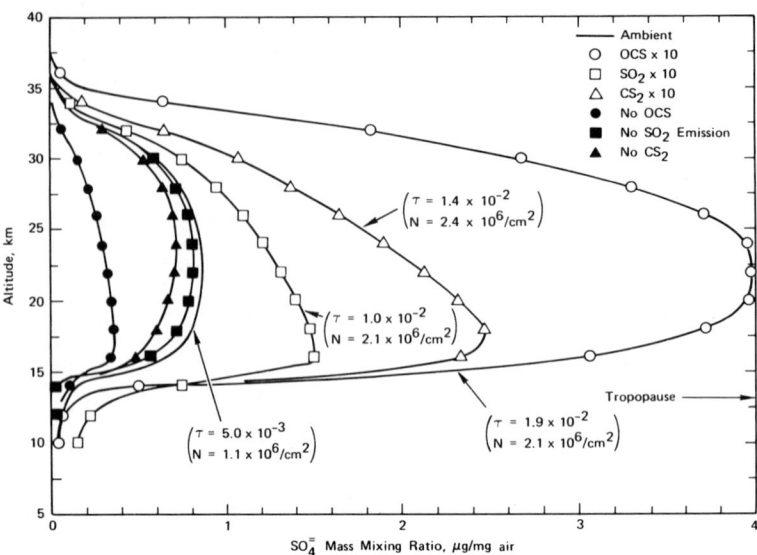

Fig.4.16. Calculated stratospheric aerosol sulfate mass mixing ratios corresponding to ambient and perturbed background sulfur gas levels. Tenfold enhancements in OCS, SO_2, and CS_2 were simulated by increasing their surface concentrations tenfold, resulting in roughly a factor of 10 increase in their surface emission rate and tropopause concentration. The abundance of only one constituent was altered for each perturbation run. For several of the cases, the total large particle ($r > 0.15$ μm) column concentration in the stratosphere (N), and the optical depth of the sulfate layer at 550 nm (τ), are given in brackets. [4.30]

burden of OCS. TURCO et al. [4.30] estimated the possible impact of industrial sulfide emissions on stratospheric aerosols and climate, assuming an eventual tenfold increase in atmospheric OCS and CS_2 concentrations by the middle of the next century. The predicted sulfate aerosol mass mixing ratio profiles are shown in Fig. 4.16.

A tenfold increase in OCS due to anthropogenic emissions might cause a fourfold increase in the mass and optical depth of the aerosol layer. TURCO et al. [4.30] estimated that the additional opacity could lead to a global surface cooling of nearly 0.1 K, which is climatically significant.

A tenfold increase in CS_2 had a smaller effect on the aerosols, even though CS_2 carries two sulfur atoms per molecule. Most of the CS_2 reaching the stratosphere is photochemically decomposed below 16 km, and the sulfate aerosols produced have a lifetime $\ll 1$ year. OCS, by contrast, is decomposed mainly above 20 km, and the aerosols generated have a lifetime exceeding 1 year. Thus, with regard to effects on the stratospheric aerosol layer, OCS seems to be a potentially more harmful by-product of fossil fuel consumption than CS_2.

4.6 Conclusions

The mathematical basis for all comprehensive stratospheric aerosol models — the generalized aerosol continuity equation (GACE) — has been reviewed in this chapter. Adjunct equations describing particle transport and charging, among other effects, have also been discussed. Detailed procedures to digitalize the GACE for computer analysis, however, are left to more exhaustive treatises. The application of aerosol models to practical atmospheric and environmental problems, as revealed in the literature, are surveyed in Sects.4.3-5. The numerous examples cited reveal some interesting facts about generalized aerosol simulations. For example, fundamental physicochemical models derived from a GACE can be usefully applied to a wide range of aerosol systems including dusts, smokes, photochemical hazes, and tenuous clouds and fogs. Such models, moreover, are not Earth-bound, but can be used to study the aerosols and clouds on other planets and moons as well. Generalized aerosol models also lend themselves to the unification of diverse data sets and to the guidance of field experiments. Perhaps the most important application of aerosol models in the future will involve the assessment of the impacts of man's activities on the terrestrial clouds and hazes that are an integral part of our environment.

References

4.1 C.E. Junge, C.W. Chagnon, J.E. Manson: J. Meteorol. *18*, 81-108 (1961)
4.2 J.W. Burgmeier, I.H. Blifford: Water, Air, Soil Pollut. *5*, 133-147 (1975)
4.3 M.A. Kritz: "Formation Mechanism of the Stratospheric Aerosols"; Ph.D. Dissertation, Yale University (1975)
4.4 J.M. Rosen, D.J. Hofmann, S.P. Singh: J. Atmos. Sci. *35*, 1304-1313 (1978)
4.5 O. Uchino, M. Hirono, M. Maeda: J. Geomagn. Geoelectr. *32*, 365-387 (1980)
4.6 R.P. Turco, P. Hamill, O.B. Toon, R.C. Whitten, C.S. Kiang: J. Atmos. Sci. *36*, 699-717 (1979)
4.7 O.B. Toon, R.P. Turco, P. Hamill, C.S. Kiang, R.C. Whitten: J. Atmos. Sci. *36*, 718-736 (1979)
4.8 S.K. Friedlander: *Smoke, Dust and Haze* (Wiley, New York 1977)
4.9 R.C. Whitten, O.B. Toon, R.P. Turco: Pure Appl. Geophys. *118*, 86-127 (1980)
4.10 M. Hirono, M. Fujiwara, T. Itabe: J. Geophys. Res. *81*, 1593-1600 (1976)
4.11 R.P. Turco, C.F. Sechrist, Jr.: Radio Sci. *7*, 725-737 (1972)
4.12 R. Parthasarathy: J. Geophys. Res. *81*, 2392-2396 (1976)
4.13 R.P. Turco, O.B. Toon, R.C. Whitten, R.G. Keesee, D. Hollenbach: Submitted Planet. Space Sci. (1981)
4.14 N.A. Fuchs, A.G. Sutugin: In *Highly Dispersed Aerosols*, ed. by G.M. Hidy, J.R. Brock (Pergamon, Oxford 1971) pp.1-60
4.15 N.A. Fuchs: *The Mechanics of Aerosols* (Pergamon, New York 1964)
4.16 G.M. Hidy, J.R. Brock: *The Dynamics of Aerocolloidal Systems* (Pergamon, New York 1970)
4.17 S. Twomey: *Atmospheric Aerosols* (Elsevier, New York 1977)
4.18 R.D. Cadle, C.S. Kiang, J.F. Louis: J. Geophys. Res. *81*, 3125-3132 (1976)
4.19 R.D. Cadle, F.G. Fernald, C.L. Frush: J. Geophys. Res. *82*, 1783-1786 (1977)
4.20 R.P. Turco, P. Hamill, O.B. Toon, R.C. Whitten, C.S. Kiang: "The NASA-Ames Research Center Stratospheric Aerosol Model. I. Physical Processes and Computational Analogs"; NASA Tech. Paper 1362 (1979)

4.21 O.B. Toon, R.P. Turco, P. Hamill, C.S. Kiang, R.C. Whitten: "The NASA-Ames Research Center Stratospheric Aerosol Model. II. Sensitivity Studies and Comparison with Observations", NASA Tech. Paper 1363 (1979)
4.22 P. Hamill, C.S. Kiang, R.D. Cadle: J. Atmos. Sci. 34, 150-162 (1977)
4.23 P. Hamill, O.B. Toon, C.S. Kiang: J. Atmos. Sci. 34, 1104-1119 (1977)
4.24 R.D. Cadle, C.S. Kiang: Rev. Geophys. Space Phys. 15, 195-202 (1977)
4.25 D.M. Hunten, R.P. Turco, O.B. Toon: J. Atmos. Sci. 37, 1342-1357 (1980)
4.26 R.P. Turco, O.B. Toon, P. Hamill, R.C. Whitten: J. Geophys. Res. 86, 1113-1128 (1981)
4.27 R.P. Turco, O.B. Toon, J.B. Pollack, R.C. Whitten, I.G. Poppoff, P. Hamill: J. Appl. Meteorol. 19, 78-89 (1980)
4.28 P. Hamill, R.P. Turco, O.B. Toon, C.S. Kiang, R.C. Whitten: "On the Formation of Sulfate Aerosol Particles in the Stratosphere", J. Aerosol Sci. (in press, 1982)
4.29 W.A. Hoppel: J. Rech. Atm. 9, 167-180 (1976)
4.30 R.P. Turco, R.C. Whitten, O.B. Toon, J.B. Pollack, P. Hamill: Nature London 283, 283-286 (1980)
4.31 R.P. Turco, R.C. Whitten, O.B. Toon, J.B. Pollack, P. Hamill: "Carbonyl Sulfide, Stratospheric Aerosols and Terrestrial Climate", in *Environmental and Climatic Impact of Coal Utilization*, ed. by J.J. Singh, A. Deepak (Academic, New York 1980) pp.331-356
4.32 P.J. Maroulis, A.R. Bandy: Geophys. Res. Lett. 7, 681-684 (1980)
4.33 P.J. Crutzen: Geophys. Res. Lett. 3, 73-76 (1976)
4.34 E.C.Y. Inn, J.F. Vedder, D. O'Hara: Geophys. Res. Lett. 8, 5-8 (1981)
4.35 W. Jaeschke, R. Schmitt, H.W. Georgii: Geophys. Res. Lett. 3, 517-519 (1976)
4.36 H.W. Georgii, F. Meixner: J. Geophys. Res. 85, 7433-7438 (1980)
4.37 P.J. Maroulis, A.L. Torres, A.B. Goldberg, A.R. Bandy: J. Geophys. Res. 85, 7345-7349 (1980)
4.38 E.C.Y. Inn, J.F. Vedder, B.J. Tyson, D. O'Hara: Geophys. Res. Lett. 6, 191-193 (1979)
4.39 W.G. Mankin, M.T. Coffey, D.W.T. Griffith, S.R. Drayson: Geophys. Res. Lett. 6, 853-856 (1979)
4.40 P.J. Maroulis, A.L. Torres, A.R. Bandy: Geophys. Res. Lett. 4, 510-512 (1977)
4.41 F.J. Sandalls, S.A. Penkett: Atmos. Environ. 11, 197-199 (1977)
4.42 A.L. Torres, P.J. Maroulis, A.B. Goldberg, A.R. Bandy: Trans. Am. Geophys. Union 59, 1082 (1978)
4.43 R.P. Turco, R.C. Whitten, O.B. Toon, E.C.Y. Inn, P. Hamill: J. Geophys. Res. 86, 1129-1139 (1981)
4.44 F. Arnold, R. Fabian: Nature London 283, 55-57 (1980)
4.45 F. Arnold, R. Fabian, W. Joos: Geophys. Res. Lett. 8, 293-296 (1981)
4.46 A.W. Castleman, Jr., I.N. Tang: J. Chem. Phys. 57, 3629-3638 (1972)
4.47 A.W. Castleman, Jr., P.M. Holland, R.G. Keesee: J. Chem. Phys. 68, 1760-1767 (1979)
4.48 G.M. Hidy, J.L. Katz, P. Mirabel: Atmos. Environ. 12, 887-892 (1978)
4.49 L.Y. Chan, V.A. Mohnen: J. Aerosol. Sci. 11, 35-45 (1980)
4.50 J.P. Friend, R.A. Barnes, R.M. Vasta: J. Phys. Chem. 84, 2423-2436 (1980)
4.51 F. Arnold: Nature London 284, 610-611 (1980)
4.52 D.J. Hofmann, J.M. Rosen: Geophys. Res. Lett. 5, 511-514 (1978)
4.53 K.H. Käselau, P. Fabian, H. Röhrs: Pure Appl. Geophys. 112, 877-885 (1974)
4.54 R.P. Turco, P. Hamill, O.B. Toon, R.C. Whitten, R.G. Keesee: "Ion Processes Contributing to Stratospheric Aerosol Formation", Proc. 6th Int. Conf. Atm. Electricity, Manchester, U.K. 1980, paper I-9
4.55 A.L. Lazrus, B.W. Gandrud: J. Geophys. Res. 79, 3424-3431 (1974)
4.56 A.L. Lazrus, B.W. Gandrud: Geophys. Res. Lett. 4, 521-522 (1977)
4.57 R.G. Pinnick, J.M. Rosen, D.J. Hofmann: J. Atmos. Sci. 33, 304-314 (1976)
4.58 D.J. Hofmann, J.M. Rosen: J. Atmos. Sci. 38, 168-181 (1981)
4.59 J.M. Rosen: J. Appl. Meteorol. 10, 1044-1046 (1971)
4.60 N.H. Farlow, D.M. Hayes, H.Y. Lem: J. Geophys. Res. 82, 4921-4929 (1977)
4.61 N.H. Farlow, K.G. Snetsinger, D.M. Hayes, H.Y. Lem, B.M. Tooper: J. Geophys. Res. 83, 6207-6211 (1978)
4.62 R.P. Turco, O.B. Toon, R.C. Whitten, R.G. Keesee, P. Hamill: "The Mt. St. Helens Eruptions of May and June 1980: Model Studies of the Physical and Chemi-

cal Processes Occurring in the Volcanic Clouds", Proc. Symp. on the Mt. St. Helens Eruption: Its Atmospheric Effects and Potential Climatic Impact, ed. by A. Deepak (Spectrum, Hampton, VA 1982)
4.63 N.H. Farlow, V.R. Oberbeck, K.G. Snetsinger, G.V. Ferry, G. Polkowski, D.M. Hayes: Science *211*, 832-834 (1981)
4.64 E.W. Barrett, R.F. Pueschel, D.L. Wellman: "Temporal Variations of Physical and Optical Properties of Mt. St. Helens Volcanic Aerosol Size Distribution", in Abstracts, 10th Laser Radar Conf., Silver Springs, MD, Oct. 6-9, 1980, pp.153-154
4.65 R.L. Chuan, D.C. Woods, M.P. McCormick: Science *211*, 830-832 (1981)
4.66 D.J. Hofmann, J.M. Rosen: Rpt. AP-63, Dept. Physics Astron., U. Wyoming, 79 pp (1981)
4.67 D.E. Brownlee: "Microparticle Studies by Sampling Techniques", in *Cosmic Dust*, ed. by J. McDonnell (Wiley, New York 1978)
4.68 B. Fogle, B. Haurwitz: Space Sci. Rev. *6*, 278-340 (1966)
4.69 O.B. Toon, N.H. Farlow: Ann. Rev. Earth Planet. Sci. *9*, 19-58 (1981)
4.70 O.A. Avaste, A.V. Fedynsky, G.M. Grechko, V.I. Sevastyanov, Ch.I. Willmann: Pure Appl. Geophys. *118*, 528-580 (1980)
4.71 G. Witt: Space Res. *9*, 157-169 (1969)
4.72 G.C. Reid: J. Atmos. Sci. *32*, 523-535 (1975)
4.73 O.B. Toon, R.P. Turco, J.B. Pollack: Icarus *43*, 260-282 (1980)
4.74 D.J. Hofmann, D.E. Carroll, J.M. Rosen: Geophys. Res. Lett. *2*, 113-116 (1975)
4.75 J.B. Pollack, O.B. Toon, A. Summers, W. Van Camp, B. Baldwin: J. Appl. Meteorol. *15*, 247-258 (1976)
4.76 D.J. Hofmann, J.M. Rosen, T.J. Pepin, R.G. Pinnick: J. Atmos. Sci. *32*, 1446-1456 (1975)
4.77 J.M. Rosen, D.J. Hofmann, J. Laby: J. Atmos. Sci. *32*, 1457-1462 (1975)

5. Stratospheric Aerosols and Climate

O. B. Toon and J. B. Pollack

With 15 Figures

Explosive volcanic eruptions inject volcanic ash and sulfur dioxide gas into the stratosphere. The ash particles quickly fall out of the stratosphere due to their large size. The sulfur dioxide gas is photochemically converted into sulfuric acid which condenses to form submicron sized particles. These sulfuric acid particles remain in the stratosphere for a few years. The particles scatter and absorb sunlight and they scatter, absorb, and emit infrared radiation. The Earth's climate is controlled by an energy balance between sunlight entering the atmosphere and terrestrial thermal-infrared radiation escaping to space. The stratospheric particles upset this balance and thereby may alter the climate.

We review the evidence showing that abnormal weather does sometimes follow single volcanic explosions. We then show that periods of intense volcanic activity occur and that these periods have cooler temperatures than nonvolcanic epochs. A discussion of the radiative properties of volcanic particles is followed by a discussion of the theory of the effects of the particles on the Earth's radiation budget. Finally, we review models of the effects on climate of volcanic injections of particles and models of the effects on climate of antropogenic alterations of the stratospheric aerosols.

5.1 Background

Stratospheric aerosols are not normally a significant component of the Earth's climate system because their optical depth is usually quite small. However, after an explosive volcanic eruption, the number of stratospheric aerosols is greatly enhanced and their optical depth can then be as large as that of tropospheric aerosols. These volcanic aerosols alter the climate by upsetting the balance between sunlight entering the atmosphere and infrared light leaving the atmosphere. Because many historical climate changes may have been caused by volcanic eruptions, it is important to have a good understanding of exactly how the changes occur. Comparisons between observed climate changes after eruptions and model simulations provide a good test of the ability of models to correctly calculate changes in climate. Aircraft and rocket flights in the stratosphere, and the release of sulfur gases by industry, may also enhance the stratospheric aerosol layer. To avoid climate changes caused by these anthropogenic emissions a good understanding of the relation between stratospheric aerosols and climate is required.

Benjamin Franklin may have been the first scientist to suggest that volcanic explosions might affect the weather. In May 1784, he wrote [5.1]:

"During several of the summer months of the year 1783, when the effects of the sun's rays to heat the Earth in these northern regions should have been the greatest, there existed a constant fog over all Europe, and great part of North America. This fog was a permanant nature; it was dry, and the rays of the sun seemed to have little effect toward dissipating it, as they easily do a moist fog arising from the water. They were, indeed, rendered so faint in passing through it that, when collected in the focus of a burning glass, they would scarce kindle brown paper. Of course, their summer effect in heating the Earth was exceedingly diminished.

Hence, the surface was early frozen.

Hence, the first snows remained on it unmelted, and received continual additions.

Hence, perhaps the winter of 1783-1784 was more severe than any that happened for many years.

The cause of this universal fog is not yet ascertained. Whether it was adventitious to this Earth, and merely a smoke proceeding from the consumption, by fire, of some of those great burning balls, or globes, which we happen to meet with in our course round the sun, and which are sometimes seen to kindle and be destroyed in passing our atmosphere, and whose smoke might be attracted and retained by our Earth; or whether it was the vast quantity of smoke, long continuing to issue during the summer from Hecla, in Iceland, and that other volcano which arose out of the sea near the island, which smoke might be spread by various winds over the northern part of the world, is yet uncertain.

It seems, however, worthy of the inquiry, whether other hard winters, recorded in history, were preceded by similar permanent and widely extended summer fogs. Because, if found to be so, men might, from such fogs, conjecture the probability of a succeeding hard winter, and of the damage to be expected by the breaking up of frozen rivers in the spring; and take such measures as are possible, and practicable, to secure themselves and effects from the mischiefs that attend the last."

Franklin noted that the "dry fog" from the volcanic eruption (or passing comet) preceded the bad winter of 1783 — a statistical connection between volcanic eruptions and climate. More importantly, he advanced a hypothesis to explain how the volcanic "fog" affected the weather. Franklin thought that the fog prevented sunlight from reaching the surface, and he was able to qualitatively demonstrate this fact with his magnifying glass. He reasoned that with less sunlight available the surface should cool off.

Since Franklin's observations many other scientists have studied the relation between volcanic explosions and weather. Most of these studies have consisted of statistical correlations between bad weather and single eruptions or between climatic anomalies and series of volcanic explosions. These studies do show that abnormal weather often occurs after large volcanic eruptions. In addition, the major climatic shifts of the past 500 years occurred in parallel with variations in the level of volcanic activity. On the other hand, however, there is no evidence that volcanic explosions preceded and initiated the ice ages. Abnormal weather occurred in many years — such as 1976 to 1978 when there were droughts in the western United States and heavy snows in the eastern United States — without evidence of any major volcanic explosions. Volcanic eruptions are not responsible for all climate changes, nor all years of bad weather, but they may cause some of them.

These statistical studies have been augmented by theoretical calculations of the effects of volcanic eruptions on climate and by measurements of the properties of volcanic debris. We now understand some ways in which volcanoes affect the weather. Benjamin Franklin's basic idea has been made quantitative, and the magnitude of the weather and climate changes observed has been found to be consistent with the theory. Of all the potential causes of climate change, volcanic eruptions are the best understood and best documented. This understanding, and the fact that the volcanic eruption establishes a unique time marker to identify when the perturbation occurs, makes volcanic explosions attractive for testing the ability of models to calculate climate changes correctly.

In the following sections the statistical evidence for a relation between volcanic explosions and climate changes is described in detail. In addition, the observed properties of volcanic debris are presented, the mechanisms by which the explosions affect climate are reviewed, and the theoretical models of the link between volcanic eruptions and climate are discussed. Finally the climatic implications of anthropogenic perturbations to the stratospheric aerosol layer are reviewed.

5.2 Statistical Relations Between Volcanic Explosions and Climatic Changes

5.2.1 Changes Observed After Single Eruptions

Mt. Agung erupted in March 1963, killing over 1500 people and creating what *National Geographic* called a "Disaster in Paradise" [5.2]. This eruption is the only one that has occurred since 1912 that had a magnitude large enough to have a significant effect on the transmission of light through the atmosphere because of the enhancement of the stratospheric aerosol mass. The debris from the eruption and its direct effects on the transmission of light through the atmosphere were mainly restricted to the southern hemisphere. Stratospheric temperatures increased after the eruption of Mt. Agung, while tropical surface temperatures decreased.

Figure 5.1 presents stratospheric temperatures during the period of the Mt. Agung eruption from radiosonde flights at a range of latitudes [5.3]. The temperature of the stratosphere normally undergoes a quasi-biennial oscillation with an amplitude of about $1°C$. In the northern hemisphere, the oscillation showed no marked change at the time of the Mt. Agung eruption (noted by an arrow). Very close to the equator the stratospheric temperature increased after the Mt. Agung eruption, but the

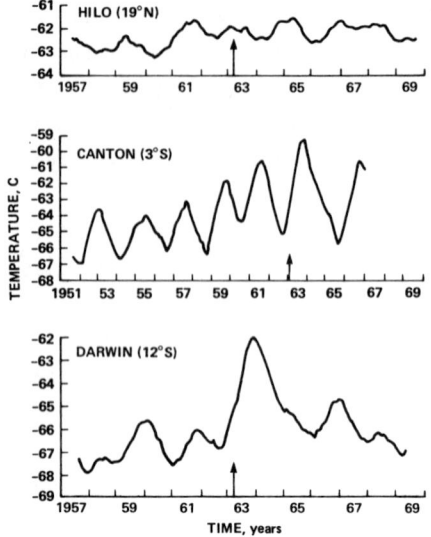

Fig.5.1. The temperature of the stratosphere at an altitude of about 20 km during the period of the Mt. Agung eruption. Notice that only in the southern hemisphere can an unambiguous temperature increase be observed. The dust from Mt. Agung was mainly restricted to the southern hemisphere. The arrow denotes the time of the Mt. Agung eruption [5.3]

amplitude of the quasi-biennial oscillation had been steadily increasing for several years, and the increased temperature could have been part of that trend. However, in the southern hemisphere, where the quasi-biennial oscillation had rather constant amplitude, stratospheric temperatures did increase dramatically after the Mt. Agung eruption. These trends are consistent with the observed spread of the dust. Very little dust entered the northern hemisphere, whereas heavy concentrations of volcanic dust were found in the southern hemisphere [5.4]. As will be shown, the magnitude of the warming observed after the Mt. Agung eruption is consistent with theoretical calculations. The warming is due mainly to the absorption of terrestrial infrared energy by the volcanic aerosols, with a smaller contribution due to absorption of sunlight by the volcanic aerosols in the stratosphere.

Changes in tropospheric temperatures during the period of the Mt. Agung eruption were discussed by ANGELL and KORSHOVER [5.5]. Figure 5.2 presents the departures of the yearly mean temperature from the long-term mean at the surface and between 850 and 300 mb for the northern hemisphere extratropics ($30°N - 90°N$), the tropics ($30°S - 30°S$), and the southern hemisphere extratropics ($30°S - 90°S$). In the troposphere and at the surface a cooling occurred worldwide after the eruption of Mt. Agung. Although the temperature decreases at the surfaces are larger than the standard deviations of the data, and they are correlated among the three latitude regions, the magnitudes of the temperature changes are not unusual. It is interesting to note that the largest temperature decrease at the surface ($0.6°C$) after the Mt. Agung eruption occurred in the northern hemisphere and therefore could not have been caused directly by the volcanic dust. However, it should be noted that the radiation field is coupled to the atmospheric dynamics in a complicated manner and, therefore, climatic changes in regions which are not directly exposed to the radiative per-

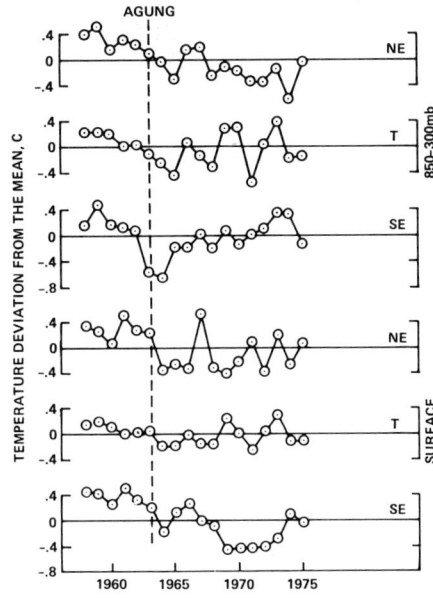

Fig.5.2. Time variation of the yearly temperature in the northern hemisphere extratropics (NE), tropics (T), and southern hemisphere extratropics (SE) at two heights during the period of the Mt. Agung eruption. The dashed line denotes the year of the Mt. Agung eruption. [5.5]

turbation are clearly possible. The atmospheric cooling was greatest (0.7°C) in the southern hemisphere extratropics.

The Mt. Agung eruption is the only major one for which stratospheric temperatures were observed. However, surface-temperature records and indirect measures of surface temperature are available for several centuries. Several studies have been made to determine whether or not a surface cooling, such as that observed after the Mt. Agung eruption, occurred after all eruptions. Numerous scientists, beginning with HUMPHREYS [5.1] found that a surface cooling does generally occur after large eruptions. LAMB [5.6] felt that the cooling was so deterministic that he used it as one indicator to develop his dust veil index to rate the magnitudes of eruptions. For example, he ascribed a hemispheric cooling of 0.4°C to Mt. Agung, 0.3°C to the 1912 eruption of Mt. Katmai, 0.6°C to the sequence of eruptions occurring from 1902 to 1904, and 0.5°C to the 1883 eruption of Mt. Krakatau. BRAY [5.7], MASS and SCHNEIDER [5.8], and MILES and GILDERSLEEVES [5.9] have made statistical studies of the temperature changes observed after the largest eruptions since 1811. An example from MASS and SCHNEIDER [5.8] is shown in Fig.5.3. They found an average temperature decrease of 0.3°C in the year after the eruption and 0.1°C or more two years after the eruption. In only a few cases did the temperatures increase after a major eruption. All the studies concluded that there was only about a 1% chance that the cooling after the eruptions was accidental and was not the result of the eruption.

The climate changes that occurred after individual eruptions may seem to be small, but these hemispheric temperature changes are often accompanied by much

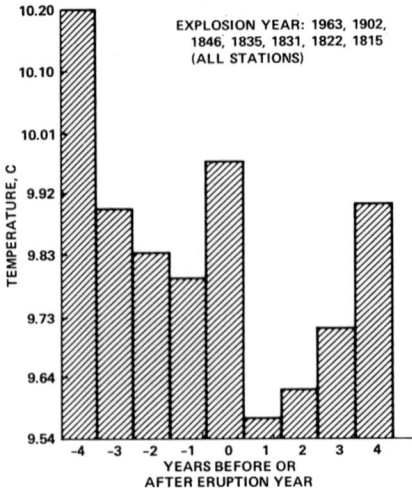

Fig.5.3. Mean temperature at a large number of stations during the years before and after several large eruptions. A statistically significant decrease of about 0.3°C occurs during the year following an eruption [5.8]

larger local variations. For example, ARAKAWA et al. [5.10] found that bad harvests in Japan prior to 1912 occurred frequently after large eruptions because of cool summers and early frosts. LAMB [5.11] stated that, over a span of many centuries, all the years of very early fall frosts (as recorded by frost-damaged rings of trees in the Swiss Alps and in California) were years following major volcanic eruptions.

Perhaps the most famous example of a volcanic eruption that had a profound effect on weather is that of Mt. Tambora. The Mt. Tambora eruption of 1815 killed about 50,000 people in Indonesia in the most violent eruption of recorded history. The following year became known as "the year without summer", or "eighteen-hundred and froze to death." Average summer temperatures were 1° to 2.5°C cooler than normal throughout New England and Western Europe [5.12]. In addition, the general cooling was accompanied in New England by snow during June, a frost in mid-July, and then an early fall frost beginning in late August [5.13,14]. The frost killed crops in New England, the price of corn increased, and the winter of 1816 was one of great hardship. Many people moved to warmer regions of the United States after 1816, adding to the general migration from New England during the early 1800s. It was likewise cold in Great Britain, France, and Germany during 1816. Europe was just beginning its recovery from the Napoleonic Wars; the Battle of Waterloo was fought during 1815. The bad weather led to crop failure, famines, and riots. The price of wheat in France was higher during 1817 than at any other time during the period from 1801 to 1912.

LANDSBERG and ALBERT [5.12] have examined the statistical significance of the summer weather of 1816. Examining eleven stations, they found that the average summer temperature in 1816 was depressed by only one or two standard deviations from the mean summer temperature at each station and that most of the eleven sites had their coldest summers in years other than 1816. On the basis of these findings, they suggested that the harsh summer of 1816 could simply have occurred by chance. Their analysis, however, suffers from two defects. First, the decrease in the mean temperature was not the critical factor in the harsh summer of 1816. The critical factor was rather a series of cold storms with attendent snow and frost during the summer. A better criterion for comparison of the summer of 1816 with other summers would have been the number of other years with summer frosts. Second, although only two of the eleven stations had their coldest summers in 1816, the coldest summers of all the other stations were in years during which the weather may have been volcanically perturbed. Two sites had coldest summers in 1913, after the 1912 Mt. Katmai eruption, and one sites was coldest during 1832, after the eruptions of Mt. Giulia and Mt. Babuyan in 1831. Four sites had their coldest summers in the years 1840, 1830, 1764, and 1732; LAMB [5.6] believed that the weather during these years was volcanically perturbed (partly because of the low temperatures during these years), but he did not have direct evidence of stratospheric dust. The final two sites had their coldest temperatures in 1929 and 1953, which are not usually believed to have been years following volcanic eruptions. However, optical-depth records do show small to moderate stratospheric aerosol perturbations in both these years [5.6,15]. The years 1928, 1932, and 1953 are the only ones between 1912 and 1963 which have been suggested as having minor volcanic perturbations. Hence, the LANDSBERG and ALBERT [5.12] list of years of lowest summer temperatures includes five sites with coldest temperatures following major eruptions, four sites with coldest temperatures during periods with possible major eruptions, and two sites with coldest temperatures following the only minor perturbations suggested between 1912 and 1963.

5.2.2 Changes Observed During Epochs of Volcanic Activity

Numerous scientists, noting that single volcanic eruptions have been followed by cooler weather during the year or two after the eruption, have attempted to correlate periods of intense volcanic activity with epochs of cold weather. Closely spaced multiple volcanic explosions are needed to cause climate changes over time scales of decades or longer because aerosols generated by a single eruption remain in the stratosphere for only a few years. Volcanoes were once considered a prime candidate for having initiated ice ages. However, this idea has not stood the test of time. On the other hand, evidence has been obtained which suggests that the climatic shifts of the last few hundred years have occurred in parallel with variations in the level of volcanic activity.

The volcanic theory of ice ages has suffered a decline in popularity for two reasons. First, the ice-age record has greatly improved. It is now clear that there have been a large number of glacial advances during the past few million years, and these glacial advances have been highly periodic. The periods of the ice ages correspond well with the characteristic periods of variations in some of the earth's orbital and axial parameters [5.16], but there is no evidence that volcanic activity is highly periodic. Secondly, a detailed record of volcanic activity during the previous glacial advance shows that there was no increase in volcanic activity at the beginning of the glacial advance.

The record of volcanic activity during the last few million years comes partly from ocean-bottom cores and partly from ice cores. KENNETT and THUNNELL [5.17] counted the number of volcanic ash layers as a function of time in ocean cores and presented evidence that the level of explosive volcanic activity had increased dramatically during the Quaternary when the glacial advance occurred (Fig.5.4). However, these data are not sufficient to determine whether or not the volcanic activity caused the ice advance. It is even possible that the large changes in the mass of ice on the continents might have caused a variation in the level of volcanic activity. The volcanic aerosols might then have had an effect in making the ice ages more severe even though they did not cause the ice ages.

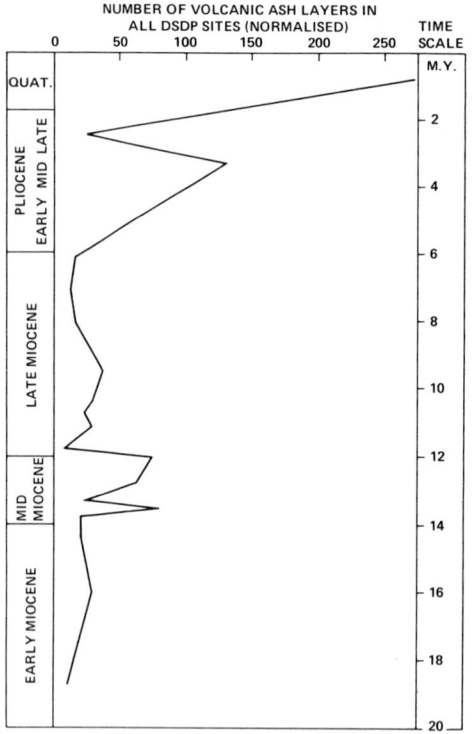

Fig.5.4. The record of explosive volcanic activity during the past twenty million years. During the Quaternary ice ages the level of volcanic activity increased (from [5.17], copyright 1977 by the American Association for the Advancement of Science)

During the Wisconsin glacial cycle, which occurred over the past 100,000 years, records of climate and volcanic activity were preserved in the Greenland and Antarctic polar caps. The oxygen isotope ratio (18/16) provides a climate indicator (it is mostly related to ice-sheet volume), while the abundance of volcanic silicate particles and sulfate concentrations provides a measure of the level of volcanic activity. Ice cores from both poles show no enhancement of volcanic particles

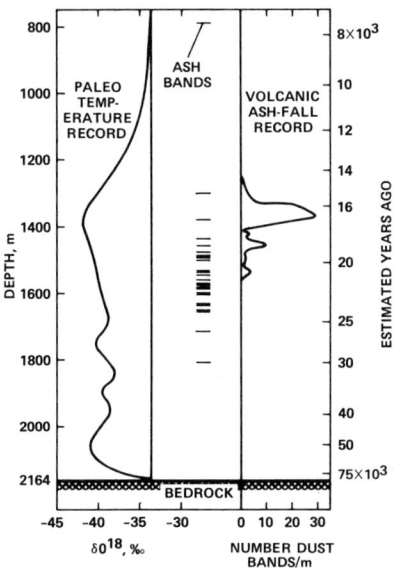

Fig.5.5. The record of explosive volcanic activity during the past 75,000 years as recorded in Antarctic ice. Note that many eruptions occurred at the maximum extent of the ice age 15,000 years ago, but none is recorded at the beginning of the ice age. [5.18]

during the early stages of the Wisconsin glacial cycle, indicating that an epoch of volcanic activity did not trigger the ice age (Fig.5.5) [5.18]. Although there is a large increase in particle concentrations in the Greenland core when oxygen isotope ratios show that the glacial maximum had been reached, the particles are not of volcanic origin but rather are windblown dust probably created by the North American ice sheets [5.19]. In Antarctica (Fig.5.5) the ice sheets do record a remarkable amount of explosive volcanic activity over a 15,000-year period at the maximum of the Wisconsin ice age [5.18]. Volcanoes may have affected the weather at the maximum of the ice age, but currently there is no evidence that they initiated the Wisconsin ice age.

The record of volcanic activity during the previous several hundred years has been statistically compared with the record of climate by a large number of researchers. The meaning of these comparisons obviously depends on the reliability of the climate and volcanic-activity records.

The record of volcanic activity comes from a variety of sources. LAMB [5.6], in his classic paper, attempted to identify all the major volcanic eruptions since 1500 from geologic records, direct observations of the explosions, or indirect observations of atmospheric phenomena. He then rated the importance of the eruption according to his dust veil index by using the amount of material ejected by the volcano, the temperature decrease observed after the eruption, or, after 1880, the decrease in transmission of sunlight observed after the volcanic eruption. As a standard, Lamb rated the Mt. Krakatau eruption as 1000 on the basis of all three criteria. Only the transmission criterion is a direct, independent measure of the magnitude of the eruption with respect to its potential effect on climate. Clearly Lamb's method of rating volcanic activity is somewhat subjective. Mt. St. Helens provides an example of an eruption which, if it had occurred before 1880, could have been improperly assessed by Lamb's index. The amount of material ejected by Mt. St. Helens would have led to a rating of a few hundred, but the transmission decrease would have led to a rating of much less than 100. Likewise, the dust veil index of Mt. Agung has been rated between 200 and 800 by various workers.

An alternate method of determining the level of volcanic activity is to search polar-ice cores for changing levels of volcanic debris, mainly sulfates. There are some difficulties with the polar-core record: in order to be detected the volcanic sulfate must exceed the background level of natural sulfate in the atmosphere; and local volcanic activity may be weighted above more remote activity. HAMMER [5.20] showed that the polar-core and the historical-reconstruction techniques of assessing the volcanic record since 1770 do differ in detail. Both records, however, show that very few explosive volcanic eruptions have occurred since 1912, but prior to 1912 there were many eruptions of significant magnitude. In addition, LAMB [5.6] found that the most heavily perturbed periods since 1500 were from about 1750 to 1850 and from about 1880 to 1900. It would be useful to extend the polar-core data to the period before 1770.

Recently geologists have attempted to construct more complete lists of important volcanic eruptions than the one constructed by LAMB [5.6]. Unfortunately these attempts, while yielding large numbers of eruptions overlooked by LAMB [5.6], can be very misleading. Only very special volcanic events actually contribute enough material to the stratospheric aerosol layer to affect the climate. Not only must the eruption be violently explosive, it must also be sulfur-rich. Most of the stratospheric aerosols that remain in the stratosphere for a long period after an eruption are composed of sulfuric acid rather than volcanic ash.

The list of volcanic explosions of "large magnitude" discussed by BRYSON and GOODMAN [5.21] shows as many eruptions during the late 1960s and early 1970s as occurred during the period near 1900. However, optical-depth records (Fig.5.6) [5.22] clearly show that the optical depth during the late 1960s and early 1970s was not strongly perturbed. Many of the "large-magnitude" eruptions they considered either were not explosive enough to inject large quantities of sulfur gas into the stratosphere or did not contain large quantities of sulfur gases in their magma. Likewise, SELF et al. [5.23], who considered only the volume of ejecta, rated the 1963 Mt. Agung eruption as only of modest significance while they rate the eruptions of Bezymjannaja (1956) and Quizopu (1932) as very large and important. These latter two eruptions, although undoubtedly of great explosive magnitudes, were probably poor in sulfur gases because their optical-depth perturbations were quite small (≤ 0.02). These geologic lists of important volcanic eruptions will be more useful in the future when it may be possible to relate the volume of sulfur injected into the stratosphere to the volume of ash removed from the volcanic crater for a particular type of eruption.

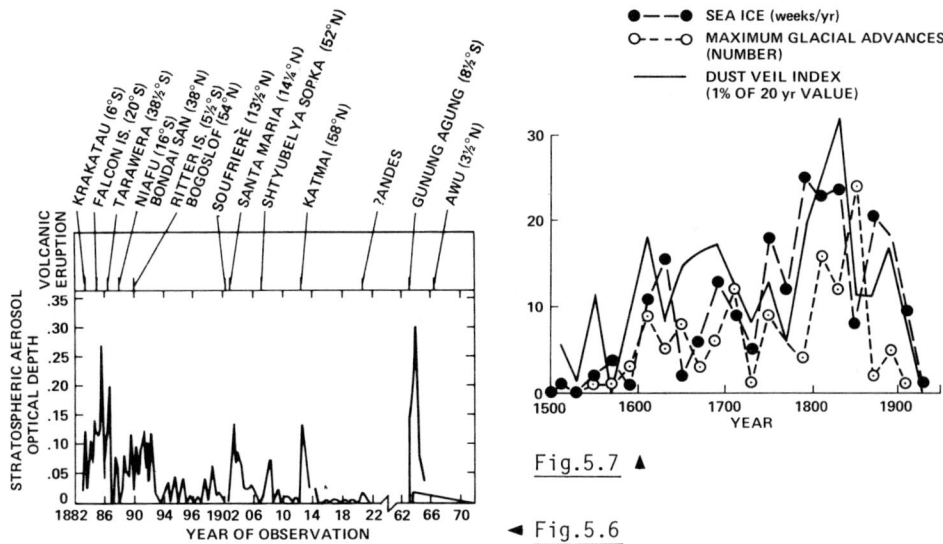

Fig.5.6. Volcanic optical depths and notable volcanic eruptions since 1880 are recorded. Note that a single eruption may yield an optical depth in excess of 0.1 for a year or two after the eruption. Volcanic activity decreased after 1912. The eruption of Mt. Agung is the only large eruption to have occurred since 1912. Although large optical depths occurred in the southern hemisphere after the Mt. Agung eruption, only small optical depths occurred in the northern hemisphere, as shown by the double curve after 1963. [5.22]

Fig.5.7. A record of volcanic activity during the past 500 years is compared with two climate indicators. The volcanic activity is measured by LAMB's [5.6] dust veil index (excluding values based solely on the temperature record). The figure uses 20-year summations, which must be multiplied by 100 to obtain the actual values in Lamb's tables. The number of weeks per year of Icelandic sea ice averaged over 20-year intervals is from [5.25]. The number of maximum glacial advances at 20-year intervals are compiled in [5.24]. Note that the record of volcanic activity and ice extent compare favorably in detail. Each period of enhanced volcanic activity corresponds to a period of ice advance, and the overall magnitudes of the cycles match. These data suggest that the coldest portion of the little ice age was during the early 1800s

The climate record prior to 1900 is not well known. Three types of climate records have been compared with the volcanic record, as well as with the record of solar activity, which has long been suspected to have an affect on climate.

First, LAMB [5.6] and BRAY [5.24] showed that the maximum level of volcanic activity, as measured by the dust veil index (and excluding values based solely on the temperature record), coincided with the maximum extent of sea ice near Iceland and the maximum number of glacial advances in the northern hemisphere (Fig.5.7) [5.25]. These maxima occurred about 1800. Presumably sea ice and glaciers are good integrators of shifts in climate over periods of many years.

Secondly, LAMB [5.6], SCHNEIDER and MASS [5.26], and ROBOCK [5.27,28] among many others since the beginning of this century have tried to correlate the record of volcanic activity with records of short-response-time climate indicators extending back to about 1600. LAMB, as well as SCHNEIDER and MASS, referred primarily to high-northern-latitude temperature records, especially those from England, and found that a cool period in the early nineteenth century and the warming at the beginning of the twentieth century correlate well with the volcanic record. However, these authors feel that the coldest period occurred during the middle of the seventeenth century, which corresponds to the Maunder minimum in sunspots. ROBOCK [5.28], on the other hand, studied a new temperature curve that is supposed to represent the entire northern hemisphere back to 1600. This newer curve does not show the seventeenth century to have been abnormally cool and therefore shows no correlation with the Maunder minimum. Indeed, these hemispheric data suggest that the early 1800s were the coldest period of the little ice age. ROBOCK [5.28] concluded that volcanic activity correlates well with the temperature records of the past 400 years, whereas solar activity does not. More work on the history of climate during the past several hundred years is needed to better determine which climate forcing, volcanic or solar, is most significant for climate.

Another comparison between volcanic activity and climate has been based on the warming trend between the late 1800s and the mid-1900s that correlates well with the decline in volcanic activity. The importance of this correlation is that both the temperature and the volcanic-activity are relatively well known during this period. Although many investigators agree that the correlation is quite good, one major point of argument has been the time lag between the volcanic activity and the climate variation. The volcanic activity ceased about 1912, but temperatures did not reach their maximum until about 1940. MILES and GILDERSLEEVES [5.9] felt that this delay was too long to indicate a correlation, but most other investigators, including ROBOCK [5.27], thought that the delay was reasonable. An additional problem is that the northern hemisphere cooling that occurred after 1940 does not correlate with an increase in volcanic activity, and, therefore, at least the most recent portion of the climate record has not been controlled by volcanoes.

Statistical studies of the relations between climate and volcanic activity do show that large volcanic eruptions have an effect on the hemispheric mean temperature in the year following the eruption. Furthermore, the level of volcanic activity correlates better than any other factor with the envelope of climate variability during the previous 500 years. An important point that is often overlooked is that the link between volcanoes and climate is not just statistical. Unlike other potential causes of climate variability, such as solar luminosity fluctuations, the effects of volcanoes on climate can be quantified because their effect on the Earth's radiation balance can be measured.

5.3 Theoretical Relationships Between Volcanic Explosions and Climate

The Earth's climate is controlled by a balance between solar energy, which is absorbed by the atmosphere and surface, and infrared energy, which is radiated back to space by the Earth's surface and atmosphere. This balance can be upset by the particles that large volcanic explosions inject into the atmosphere. The result is a time-varying vertical temperature profile that acts, within the time constants of the climate system, to achieve a new balance.

Unusual optical phenomena, often reported for a year or two after large eruptions, provide direct visual evidence that volcanic particles affect sunlight. A large, bright, colored ring surrounded the sun and moon after the Mt. Krakatau eruption; it was called Bishop's ring after the Rev. Bishop who first reported it. Sometimes after an eruption the sun and moon appear blue or green, perhaps giving rise to the phrase "once in a blue moon". In addition, the sky color may change from its normal deep blue to milky white. The most vivid signs that particles affect sunlight are the brilliant twilight colors that follow even minor explosive eruptions. Such colors were seen after the eruption of Volcán de Fuego in 1974 as well as after the eruption of Mt. St. Helens. There were so many eruptions in the first decades of the nineteenth century that the twilight may normally have been quite colorful. It has been suggested that the English romantic painter J.M.W. Turner recorded these "volcanic" twilights in his landscape scenes, which became famous for their colorful skies. The explanations of these colored rings and tinted skies are to be found in simple applications of the theories of light scattering and gaseous absorption in the atmosphere [5.29].

Volcanic eruptions might affect the climate in many ways other than radiative ones; for example, the particles could alter the physical properties of water clouds. However, most attention has been given to the direct effects of the particles on the radiation budget, and the present discussion will be limited to these radiative effects. The theory of the interaction of light with particles depends on several parameters. In the following sections the current knowledge of these parameters is discussed; the sensitivity of climate calculations to the values of the parameters is examined; and, finally, calculations of the effects of volcanic aerosols on climate are reviewed.

5.3.1 Radiative Properties of the Aerosols

In order to determine theoretically the net effect of the volcanic particles on both the solar and the thermal components of the radiation budget, several aerosol properties must be known; these properties are discussed in detail by TOON and POLLACK [5.30,31]. First, the quantity of aerosols must be known; it could be determined by measuring the aerosol concentration from an airplane or the optical depth from a ground station, but it is best determined from satellites such as

SAGE and SAM [5.32]. The advantage of satellites is that global coverage is obtained and contamination of the measurements by tropospheric aerosols is easily avoided. Secondly, the optical constants of the aerosols at wavelengths varying from the near ultraviolet to the far infrared must be determined. They can be determined indirectly by finding the optical constants of the chemicals that compose the aerosols. This approach works well for infrared wavelengths, but the absorption of visible light may be dominated by minor impurities. Direct laboratory measurements of collected aerosols or in situ measurements are needed in order to determine accurately the absorption of visible light. Thirdly, the wavelength dependence of the aerosol opacity must be known in order to determine the relative effects of the aerosols on visible and infrared light. This dependence is now best calculated by measuring the aerosol size distribution and using Mie scattering theory. Direct measurements could be made far into the infrared with current instruments, and these measurements would provide an interesting check on the calculations, although care would be needed to separate gas from aerosol effects at these wavelengths. Finally, the angular scattering properties of the aerosols must be determined in order to quantify the amount of forward-scattered sunlight. Presently the scattering phase function can be calculated only by using measured aerosol size distributions, but it will soon be possible to measure it by using aircraft-borne nephelometers. Measurements made after the Mt. St. Helens eruption have greatly added to our knowledge of all these properties.

Figure 5.6 presents a compilation of measurements of the stratospheric aerosol optical depth after volcanic explosions. The measurements included the entire solar spectrum. The optical depth is the logarithm of the transmission of light through the atmosphere. These measurements show that single eruptions are able to reduce the strength of the direct solar beam over large areas of the Earth by 25% for a month or two and by 10% for periods of a year or more. The visible optical depth of aerosols in the Earth's lower atmosphere is about 0.125 and, therefore, volcanic eruptions can more than double the total aerosol optical depth of the atmosphere [5.30].

Large eruptions may inject several cubic kilometers of volcanic ash into the atmosphere, most of which quickly falls to the ground [5.6]. The total volume of volcanic material in the stratosphere can be determined fairly accurately from observations of the optical depth of volcanic debris after the debris has spread over a large area of the Earth (Fig.5.6). The volume after large eruptions, such as those of Mt. Agung, Mt. Katmai, and Mt. Krakatau, ranges from 0.008 to 0.03 km^3 [5.29]. Because these volumes are only 0.1% to 1% of the total ejected volume, it might be imagined that most of the volcanic debris remaining in suspension was composed of ash. However, this idea is not correct; rather, most of the debris is apparently composed of a sulfate, probably sulfuric acid.

The evidence for the importance of sulfates is circumstantial. After the modest eruption of Volcán de Fuego in 1974, the aerosols were found to have a volatility

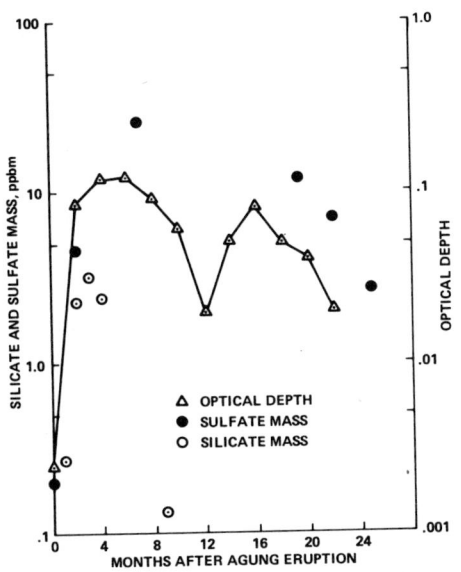

Fig.5.8. The optical depth over Australia during 1963 [5.36] is compared with the sulfate mass [5.35] and the silicate mass [5.37] measured above Australia near 20-km altitude. Note that the sulfate mass greatly exceeds the silicate mass and follows the trend of the optical depth fairly closely

similar to the normal stratospheric acid droplets [5.33]. GANDRUD and LAZRUS [5.34] measured a large increase in sulfate after the Volcán de Fuego eruption that was sufficient to account for the increased aerosol mass. After the powerful eruption of Mt. Agung in 1963, CASTLEMAN et al. [5.35] found a large increase in sulfates in the southern hemisphere stratosphere. At 40°S latitude the mass of sulfate at about 18-km altitude reached a maximum of about 30 ppbm about six months after the eruption (Fig.5.8) [5.36,37]. The equivalent hemispheric sulfate volume is very close to the 0.01 km^3 of aerosol estimated from the optical depth. Moreover, Fig.5.8 shows that the change in optical depth closely paralleled the change in sulfate mass at the same latitude. MOSSOP [5.37] made collections of stratospheric debris after the Mt. Agung eruption. Although he could not quantify the ratio of silicate to sulfate, he felt that sulfate was clearly dominant one year after the eruption. An approximate silicate mass can be obtained from Mossop's silicate data. As Fig.5.8 shows, the silicate mass is much less than the sulfate mass a few months after the eruption.

The Mt. St. Helens eruption is the first one for which the silicate-sulfate ratio has been determined and the first one for which the gas phase and particulate sulfur have both been measured. HOBBS et al. [5.38] measured the SO_2 content of the tropospheric volcanic plume on 18 May 1980 and found only modest quantities as compared with other eruptions. INN et al. [5.39] and GANDRUD and LAZRUS [5.40] measured SO_2 and sulfate in the stratosphere. Whereas COS appears to have been the dominant precursor gas for stratospheric sulfuric acid prior to the Mt. St. Helens eruption, SO_2 was found to be dramatically increased in the few-days-old volcanic plume, and it was the dominant source of new sulfuric-acid particles. The mixing ratio of the sum of the gas phase and the particulate sulfur even in the early dense stratospheric plume was only an order of magnitude larger than that found for the whole stratosphere after the Mt. Agung eruption. Within a few days, as the plume dispersed slightly, the mass mixing ratio was comparable to measurements of CASTLEMAN et al. [5.35] after the Mt. Agung eruption. Mt. St. Helens seems to have injected a surprisingly small quantity of sulfur into the stratosphere. Hence the geologic aspects of the eruption, particularly the low sulfur content of the magma, may have played the most significant role in determining the impact of Mt. St. Helens on climate.

The silicate-sulfate ratio has been reported so far only for the first few days after the Mt. St. Helens eruption [5.41]. Four days after the eruption, the masses of ash and sulfate were comparable, but the sulfate mass was rapidly increasing because of the chemical conversion of the gas-phase sulfur and the silicate mass was rapidly decreasing as the large rock particles fell out of the stratosphere. The evolution of silicate and sulfate provides an important test for stratospheric aerosol models (Chap.4), which may allow the models to "scale" the Mt. St. Helens eruption and project the behavior of an eruption ejecting larger quantities of sulfur. The time evolution of the composition of Mt. Agung debris (Fig.5.8) and the OH concentration and limited rate of conversion of sulfur gases to sulfuric acid indicate that the relatively rapid conversion of gases to particles in Mt. St. Helens debris (~1 week) will be stretched to long time periods (months) after large sulfur-injection events.

The optical constants can be selected by utilizing this knowledge of the composition of the volcanic debris. The infrared optical constants of concentrated sulfuric-acid solutions [5.42] are probably the most appropriate for the sulfate component of the volcanic aerosols. The dust component after the Mt. St. Helens eruption was largely composed of glass [5.43], suggesting that the infrared optical constants of obsidian or basaltic glass [5.44] are probably most appropriate for acidic or basic eruptions, respectively. Typically explosive volcanic events are toward the acidic end of the silicate scale.

The mass of sulfate is probably greater than the mass of silicate in the long-lived volcanic dust veil and the visible, real refractive index of sulfuric acid probably best describes volcanic debris. However, even a small amount of silicate, if it is moderately absorbing, could control the absorption of sunlight. The visible single-scattering albedo after the Mt. St. Helens eruption was found to be 0.98 four days after the eruption and 0.995 nine days after the eruption [5.45]. These values are crudely consistent with the values expected for a mixture of silicates and sulfates if the measured visible imaginary index of silicates [5.46] is used and the imaginary index of pure sulfuric acid [5.42] is used. For example, a value of 0.98 for the single-scattering albedo implies that the silicates are responsible for about one-third of the total extinction, an implication which is consistent with the observed mass fractions.

In addition to the optical constants, the size distribution of volcanic aerosols is required to calculate the opacity as a function of the wavelength and the angular scattering pattern. At present, there are no observations available to check calculations of the wavelength dependence of the opacity. However, it is known that volcanic aerosols are strongly forward scattering and, therefore, much of the light missing from the direct solar beam (Fig.5.6) still reaches the ground as diffuse skylight (Fig.5.9). Because volcanic particles are large enough to scatter all visible colors fairly efficiently, the sky appears milky white rather than blue after large eruptions. DYER and HICKS [5.36] presented observations of the direct, diffuse, and total skylight in Australia after the Mt. Agung eruption. Although the direct solar beam was reduced by nearly 25%, the diffuse skylight doubled, and thus the total sunlight reaching the surface was only slightly diminished (Fig.5.9).

MOSSOP [5.37] measured the size distribution of volcanic ash after the eruption of Mt. Agung, and the measurements were reviewed by TOON and POLLACK [5.30]. The size distributions, which are illustrated in Fig.5.10, were basically lognormal with a time-independent standard deviation of about 1.8. During the first month after the eruption the mode radius was about 0.5 to 1 μm. The mode size uniformly decreased with time to about 0.1 μm after one year. Very similar ash size distributions were reported by FARLOW et al. [5.43] during the first month after the Mt. St. Helens eruption.

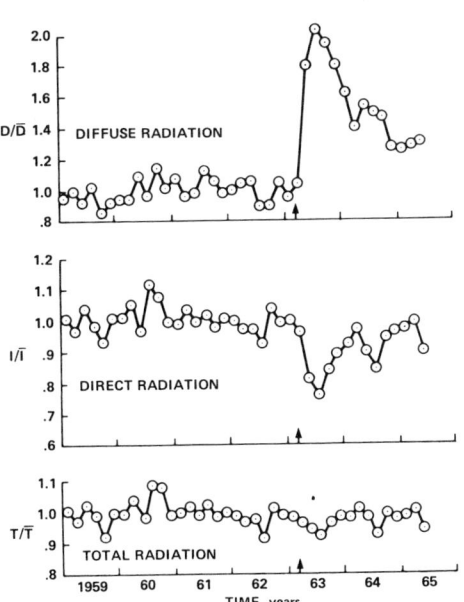

Fig.5.9. Deviations from the mean of the diffuse $D/\bar{D}$, direct $I/\bar{I}$, and total solar radiation $T/\bar{T}$ in Australia during the period of the Mt. Agung eruption (marked by arrow). Note that an increase in diffuse radiation compensates for the large decrease in direct radiation and, therefore, the change in total radiation is very small. [5.36]

The size distribution of sulfates after an eruption is not well known. TOON and POLLACK [5.30] suggested that six months after the Mt. Agung eruption the unperturbed acid aerosol size distribution was consistent with the available information, which was largely optical. After the Mt. St. Helens eruption, sulfuric-acid

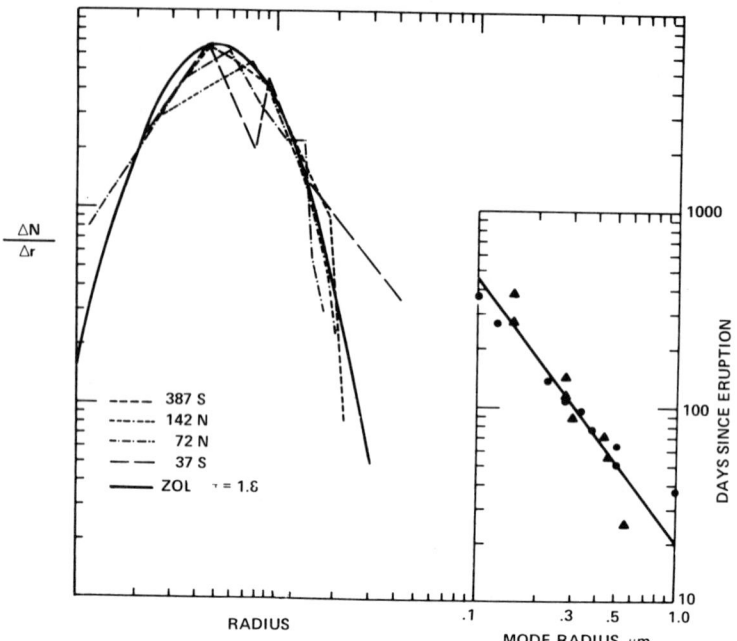

Fig.5.10. Measurements of the size distribution of volcanic ash made after the eruption of Mt. Agung [5.37]. The size distributions have been overlapped to show that they all have a similar shape and width, so the scale is arbitrary. The functions are well described by a Zold function [5.30], which is a type of lognormal function. The width of this function, σ, is constant, but the mode radius varies with time as shown in the figure. The numbers in the key refer to days after the eruption. The letter S indicates that the distribution was measured between $40°$ and $45°$ S latitude, and the letter N, between $15°$ and $35°$ S latitude. The triangles are from the S measurements and the circles from the N measurements. [5.30]

size distributions were measured. CHUAN et al. [5.47] suggested that the mode radius within the first week was no more than a factor of two larger than its unperturbed value, a result consistent with measurements made at a much later time by Farlow (private communication). In any case, the Mt. St. Helens eruption was not large enough to allow one to use the observed size distribution to represent a major eruption.

It is possible to calculate the wavelength-dependent optical depth, the phase function, and the single-scattering albedo by using the size distributions and optical constants. Relevant parameters are presented in Fig.5.11 [5.30]. If one knows the optical depth τ at a given reference wavelength (Fig.5.6), the optical depth at any other wavelength may be found by using

$$\tau_\lambda = \frac{Q_{ext_\lambda}}{Q_{ext_{ref}}} \tau_{ref} \tag{5.1}$$

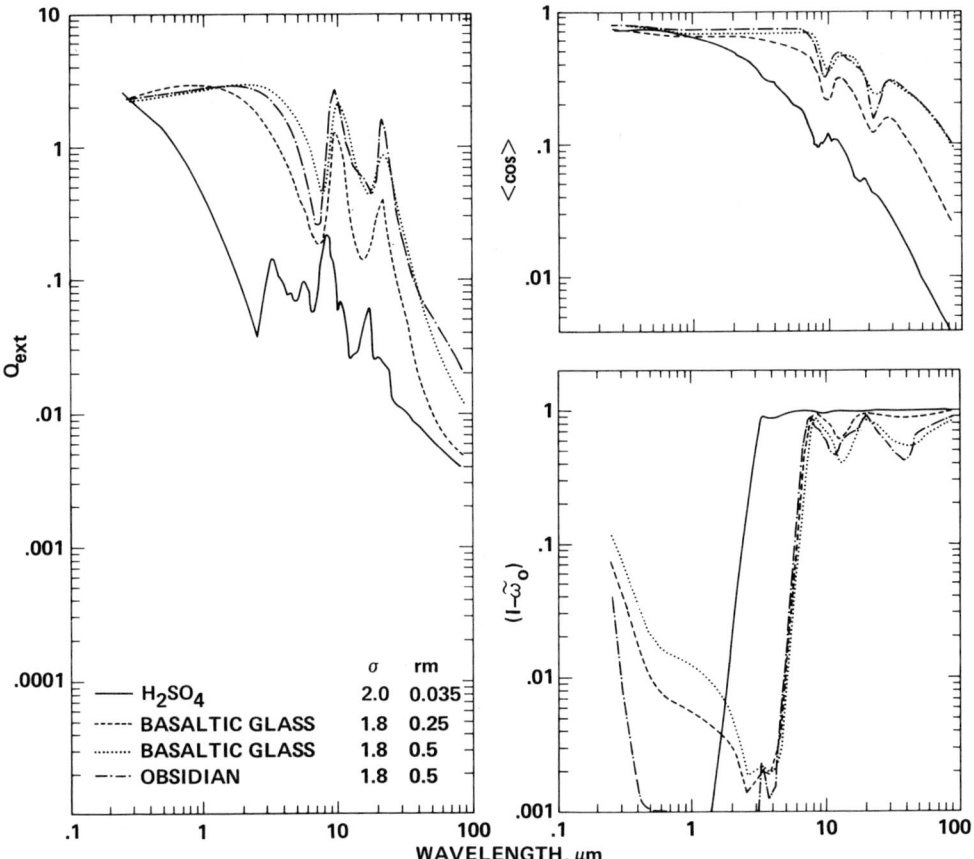

Fig.5.11. The wavelength dependence of the asymmetry factor, the single-scattering albedo, and the extinction efficiency calculated by using the appropriate size distribution for unperturbed sulfuric-acid aerosols, and by using the refractive indices of basaltic glass or obsidian in volcanic ash size distributions. [5.30]

where Q_{ext_λ} is the extinction cross section at a given wavelength. The scattering phase function $P(\cos\theta)$ is represented most simply by

$$<\cos\theta> = \int_{-1}^{1} \frac{1}{2} P(\cos\theta) \cos\theta \, d\cos\theta \quad . \qquad (5.2)$$

When the asymmetry parameter $<\cos\theta> = 1$, light is only scattered forward by the particles and, when $<\cos\theta> = 0$, light is scattered isotropically. The single-scattering albedo $\tilde{\omega}_0$ is the ratio of the scattering cross section to the extinction cross section. Particles with $\tilde{\omega}_0 = 1$ are scattering but not absorbing, whereas particles with $\tilde{\omega}_0 = 0$ are absorbing but not scattering.

The imaginary indices of refraction of the obsidian and basalt samples presented in Fig.5.11 are somewhat smaller than the values appropriate for the silicate component of Mt. St. Helens and, therefore, $\tilde{\omega}_0$ is somewhat overestimated there. For the observed size distributions of the Mt. St. Helens ash [5.43], $\tilde{\omega}_0 \simeq 0.94$ for the silicate component alone at 0.55 μm, with $\tilde{\omega}_0$ decreasing toward shorter wavelengths in a fashion qualitatively similar to that shown in Fig.5.11. The particles are also calculated to be strongly forward scattering at visible wavelengths. The large ash particles have optical depths at 10 μm that are comparable to their visible opacities. (At infrared wavelengths there is little difference between the refractive indices of various types of silicates.) The smaller sulfuric-acid particles have optical depths at 10 μm that are an order of magnitude less than their visible opacities. The assumed, unperturbed sulfuric-acid aerosol size distribution is probably a lower size limit to the volcanic sulfuric-acid particle size. Measurements and models of the volcanic sulfate aerosol size distributions, and perhaps direct measurements of the infrared opacity, are needed to improve the theory of the impact of volcanic eruptions on climate.

5.3.2 Sensitivity Studies of the Effects of Aerosols on Climate

The radiative effects of stratospheric aerosols are determined chiefly by the following parameters: their single-scattering albedo $\tilde{\omega}_0$, their asymmetry parameter $<\cos\theta>$, their area-weighted mean particle size $\bar{r}$, their optical depth in the visible τ, and the albedos of the underlying surface A_s, and atmosphere A_a [5.22,48-50]. For most purposes this list of key properties can be somewhat shortened. Except for $\bar{r}$ values much less than 0.1 μm, which are not of interest, the value of $<\cos\theta>$ in the visible is expected to deviate very little from a typical value of 0.7, and the value of $<\cos\theta>$ in the infrared is controlled by $\bar{r}$ (Fig.5.11). Also, for globally averaged situations with constant snow and ice, the value of A_s will remain constant. The change in A_a, due primarily to cloud feedback processes, is difficult to predict at the present time. Cloud amount and altitude are assumed constant for sensitivity studies. For the range of $\bar{r}$ of interest, $\tilde{\omega}_0$ is always quite small in the infrared (Fig.5.11) and hence only the visible value of $\tilde{\omega}_0$ is critical [5.51]. The value of $\bar{r}$ determines the wavelength dependence of τ and, in particular, the critical ratio of infrared-to-visible extinction coefficients. Usually a reference value of τ at visible wavelengths is used to establish the absolute value of τ at other wavelengths. The important paramters then are $\hat{\omega}_0$ and τ in the visible and $\bar{r}$.

Although many types of models have been employed to study the effects of volcanic aerosols on climate, the present focus is on one-dimensional, radiative-convective models. In these models the solar energy deposition and the infrared energy deposition are calculated throughout the atmosphere. The temperature is found by forcing the infrared energy leaving the top of the Earth's atmosphere to balance exactly the solar energy absorbed by the Earth and the atmosphere. Furthermore, the atmosphere is forced into radiative equilibrium everywhere except where the lapse rate exceeds a critical value. Larger lapse rates are observed to be prevented from occurring in the atmosphere by dynamical processes. Any effects due to heat storage in the system or dynamical redistribution of heat with latitude are ignored.

Using unperturbed stratospheric aerosol properties measured in Alaska in July 1979 from NASA Ames' U-2 aircraft and NASA's SAM II satellite, POLLACK et al. [5.51] have carried out a series of sensitivity experiments to determine the impact of the aerosols on the global radiation budget. Except where noted, the area-weighted mean particle radius was equal to about 0.1 μm. The visible optical depth was equal to 0.003, a value close to the minimum over the last several decades. Calculations of $\tilde{\omega}_0$, $<\cos\theta>$, and 0_{ext} for these choices of size distribution are illustrated by POLLACK et al. [5.51] and they are quite similar to those in Fig.5.11 for H_2SO_4.

Figure 5.12 illustrates the *change* in the Earth's spherical albedo between models with and without stratospheric aerosols as a function of their single-scattering albedo $\tilde{\omega}_0$. A critical value of $\tilde{\omega}_0$ of 0.94 separates domains where the aerosols cause an increase ($\tilde{\omega}_0 > 0.94$) or a decrease ($\tilde{\omega}_0 < 0.94$) in the spherical albedo. This critical value is insensitive to the choice of $\bar{r}$. The magnitude (but not the sign) of the albedo change is, of course, controlled by τ and scales approximately linearly with τ.

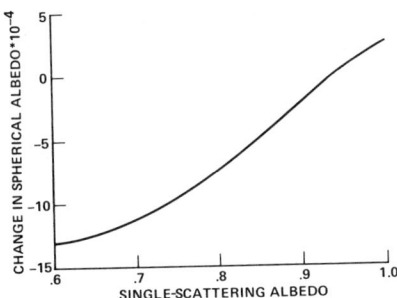

Fig.5.12. Change in the spherical albedo of Earth due to the stratospheric aerosols of July 1979 as a function of their single-scattering albedo. The optical depth at 0.55-μm wavelength was set equal to 3×10^{-3}. [5.51]

Figures 5.13a,b show the steady-state *change* in atmospheric temperatures caused by the presence of the stratospheric aerosols for various values of $\tilde{\omega}_0$ and the particle size factor F. The factor F is a constant by which the radii of the observed size distributions during July 1979 were multiplied. Size distributions with values of F as large as 10 could be maintained only for a few months because such large particles would rapidly fall out of the stratosphere. In all cases, absorption of sunlight and thermal radiation causes temperatures to increase in the portions of the lower stratosphere where the aerosols are situated. The amount of warming is quite sensitive to the values of $\tilde{\omega}_0$ and F, as has been pointed out by FIOCCO et al. [5.52] and POLLACK et al. [5.22,48]. The surface can warm or cool, depending on the values of $\tilde{\omega}_0$ and F. The sign of this change reflects the competition between cooling caused by a decrease in the amount of sunlight reaching the troposphere and warming caused by an increase in downward-directed thermal radiation from the

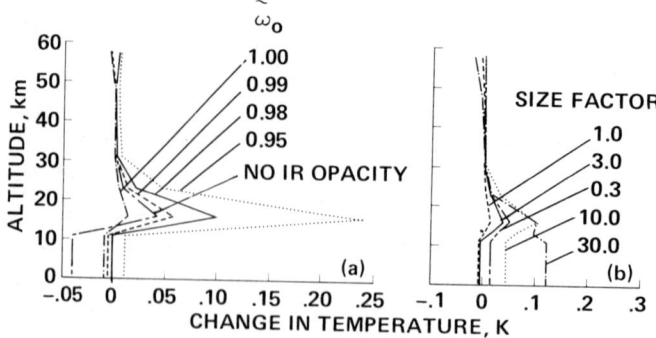

Fig.5.13a,b. Steady-state change in atmospheric temperature due to the stratospheric aerosols of July 1979 as a function of altitude. In (a) separate curves are shown for varying choices of the single-scattering albedo of the aerosols at 0.55-μm wavelength. In (b) separate curves are shown for different particle size factors as defined in the text, with the single-scattering albedo equal to 1 in all cases. A constant value of 0.003 for the optical depth of the stratospheric aerosol layer at a wavelength of 0.55 μm is used. [5.51]

stratosphere. The importance of the infrared opacity of the aerosols can be appreciated by comparing the curve labeled $\tilde{\omega}_0 = 0.99$, in which infrared opacity is included in the calculation, with the curve labeled "no infrared opacity", in which it is not included. These temperature changes are applicable when the aerosol optical depth is sustained for periods in excess of the thermal response time of the atmosphere (approximately one month for the stratosphere and approximately five years for the troposphere because of the thermal inertia of the oceans).

Values of $\tilde{\omega}_0$ in the range from 0.95 to 1.00 were measured during the Alaska flight series [5.53]. Thus, these "background" aerosols caused an increase in the Earth's spherical albedo, an increase in the stratospheric temperatures, and either a warming or a cooling of the surface. However, because their optical depth τ was only 0.003, none of these changes is climatically significant. Since the calculated changes scale approximately linearly with the optical depth of the aerosols [5.22], minimum values of 0.03 for τ are required over the response time of the atmosphere (or larger values for shorter durations) to cause a climatically significant temperature change of magnitude = 0.05 K in the troposphere and 0.5 K in the stratosphere. However, values of τ as large as 0.3 have been recorded over a year or two-year period after individual volcanic explosions, and values of τ as large as 0.1 have been recorded over a decade during times of multiple volcanic explosions (Fig.5.6). Therefore, it would seem that volcanic aerosols have been climatically significant. Values of 0.995 for $\tilde{\omega}_0$ were measured in the volcanic plumes of Mt. St. Helens during the first several weeks after its explosions in May and June 1980 [5.45]. Also, the value of F was on the order of 1.5 to 2 [5.47]. According to Fig.5.2, these values for $\tilde{\omega}_0$ and F imply that volcanic aerosols tend to cool the surface.

5.3.3 Theoretical Studies of the Effect of Volcanoes on Climate

Several investigations of the climate changes resulting from single or multiple volcanic eruptions have been conducted. In all these studies the single-scattering albedo and the particle size have been fixed at constant values. Since the single-scattering albedo observed after the Mt. St. Helens eruption [5.45] was so large, the typical choice of $\tilde{\omega}_0 \simeq 1$ is a good one. However, the size distribution varies greatly after an eruption. Soon after the initial eruption particle sizes may be large enough to cause a warming at the ground. The calculations to date have simply assumed that the size distribution several months after an eruption is the relevant one for climate time scales and that this size distribution has a value of $\bar{r}$ of a few tenths of a micron. Further observations and aerosol modeling studies of $\bar{r}$ are required to confirm these ideas.

HUMPHREYS [5.1] conducted the first theoretical studies on the climatic effects of volcanic aerosols. He argued that the volcanic particles were so small that they would backscatter to space half the solar radiation they intercepted, and that they would not affect infrared radiation. He then deduced that a cooling of $6°$ to $7°C$ would occur after a large eruption. Humphrey's basic assumptions about the large magnitude of backscattering and the negligible effects on infrared light were wrong. In fact, as has been discussed, most of the light is forward scattered, and the effects of the particles on infrared light nearly compensate for those on visible light.

HANSEN et al. [5.54] calculated the time evolution of the stratospheric and tropical tropospheric temperatures after the eruption of Mt. Agung. Figure 5.14 shows that the calculated temperatures agree favorably with the observed ones. In their model, as well as in earlier calculations by POLLACK et al. [5.22], it was found that the absorption of infrared energy by the volcanic particles was the major factor in warming the stratosphere. The troposphere cooled because the volcanic particles reflected a small amount of sunlight back to space that would otherwise have reached the surface. HANSEN et al. [5.54], as well as POLLACK et al. [5.22,48], assumed that the volcanic particles were largely sulfuric acid and that they were characterized by a size distribution that is identical to nonvolcanic size distributions [5.30]. In both calculations, the net cooling was only about half of what it would have been in the absence of infrared backwarming of the volcanic aerosols.

POLLACK et al. [5.48] calculated the warming to be expected between the late nineteenth and the middle of the twentieth centuries caused by the decline in volcanic activity, which was illustrated in Fig.5.6. The calculation was done by using the optical depths averaged over a decade to represent crudely the response time of the Earth-atmosphere system. As shown in Fig.5.15, the increase in temperature due to the decline in volcanic activity is sufficient to match the observations. As discussed previously, several workers have noted the statistical parallel between the volcanic activity and the temperature record. These calculations show that the magnitude is also correct. Of course, further work must be done before

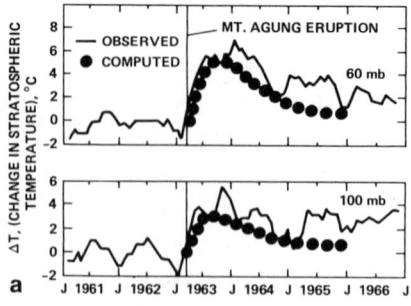 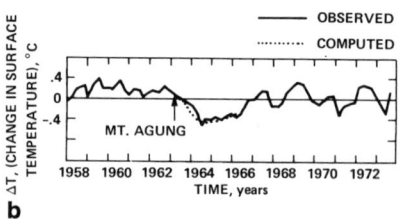

Fig.5.14a,b. Calculated temperature changes after the Mt. Agung eruption are compared with observed ones. In (a) the stratospheric temperatures increased because the aerosols absorbed terrestrial thermal radiation. In (b) the tropospheric temperatures decreased because the particles scattered sunlight back to space that would otherwise have reached the surface([5.54], copyright 1978 by the American Association for the Advancement of Science)

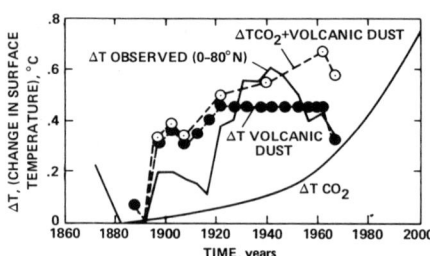

Fig.5.15. Temperature changes calculated throughout the twentieth century by using the decade average of the optical depths (Fig.5.6). The changes agree well with those observed up to 1940 and the changes due to volcanoes are much larger than the changes to be expected due to CO_2 during the first half of the century. A significant warming due to CO_2 is expected at the end of the twentieth century. [5.48]

it can be concluded that the decrease in volcanic activity was the actual cause of the warming trend between 1900 and 1940. It is interesting to note in Fig.5.15 that the increasing level of CO_2 in the atmosphere should become a significant factor in the Earth's climate sometime during the next few decades.

POLLACK et al. [5.48] discussed the role of volcanoes in several other epochs with different climates. They found that, during the little ice age between 1500 and 1900 and during the coldest period of the last glacial age 20,000 years ago, large enough optical depths probably occurred to have affected the climate significantly.

5.4 Studies of Anthropogenic Alterations of the Stratospheric Aerosol Layer

There has been some concern that man's activities could significantly enhance the "background" stratospheric aerosol layer by introducing additional sulfur-containing gases and condensation nuclei into the stratosphere. However, calculations by TURCO et al. ([5.55] and Chap.4), who used combined radiative transfer/aerosol growth models, indicate that, for anticipated traffic levels, no climatically significant perturbation to the stratospheric aerosol layer is expected to result from the effluents of supersonic transports and space shuttles flying through the stratosphere over the next several decades. Nevertheless, there is a subtle effect that warrants further study: Emissions of graphitic carbon into the stratopshere by high-flying aircraft might lower the value of $\tilde{\omega}_0$ [5.51]. Such an alteration could be important, especially during times of enhanced optical depth. TURCO et al. ([5.56] and Chap.4) have also investigated the possible effects of sulfur gases that are emitted in the troposphere and diffuse upward into the stratosphere. They find that a tenfold increase over present-day values of tropospheric COS could enhance the optical depth of the stratospheric aerosol layer to the degree that a 0.1 C decrease in $<T_s>$ would result. Thus, in the future, man's industrial activities might augment the stratospheric aerosol layer by a climatically significant amount. Obviously it is important to estimate more precisely the sources of COS and its likely growth rate in order to better determine the likelihood of its potential climatic effects.

5.5 Summary

A significant amount of evidence shows that single volcanic eruptions that inject large quantities of sulfur dioxide into the stratosphere are responsible for increasing stratospheric temperatures by several degrees and for cooling tropospheric temperatures by several tenths of a degree for a year or two after the eruption. On several occasions these small average-temperature changes have been accompanied by severe local-weather changes. Volcanic eruptions tend to occur in groups. The variations of volcanic activity during the past 500 years have occurred in parallel with changes in climate. Although there are many possible causes of climate change, including solar-luminosity changes, volcanic activity is the best-quantified agent of climate change, and it shows the greatest correlation with the climate record.

The theory of the relation between volcanoes and climate is moderately well developed. The magnitudes of calculated climate changes agree well with observed changes. However, there are several uncertain parameters in the theory, most notably the volcanic particle size distribution and the visible single-scattering albedo of the particles. New measurements of these properties after the Mt. St. Helens eruption should significantly improve the theoretical models. The theoretical models also require further development in order to consider the time and space dependence of the volcanic debris and the climate changes.

Anthropogenic changes to the aerosol layer are also possible if soot is deposited by jet aircraft in the stratosphere, or if large increases in tropospheric COS result from pollution. These possible modifications need to be monitored in the future.

References

5.1 W.J. Humphreys: *Physics of the Air* (McGraw-Hill, New York 1940)
5.2 W.P. Booth, S.W. Matthews, R.F. Sisson: Natl. Geogr. *124*, 436-458 (1963)
5.3 R.M. McInturff, A.J. Miller, J.K. Angell, J. Korshover: J. Atmos. Sci. *28*, 1304-1307 (1971)
5.4 A.J. Dyer, B.B. Hicks: Q. J. R. Meteorol. Soc. *94*, 545-554 (1968)
5.5 J.K. Angell, K. Korshover: Mon. Weather Rev. *105*, 375-385 (1977)
5.6 H.H. Lamb: Philos. Trans. R. Soc. London A*266*, 425-533 (1970)
5.7 J.R. Bray: Adv. Ecol. Res. *7*, 177 (1971)
5.8 C. Mass, S.H. Schneider: J. Atmos. Sci. *31*, 1995-2004 (1977)
5.9 M.K. Miles, P.B. Gildersleeves: Nature London *271*, 735-736 (1978)
5.10 A. Arakawa, T. Fujita, H. Itoo, Y. Masuda, S. Matsumoto, T. Murakami, T. Ozawa, T. Suzuki, M. Takeuchi, K. Tomatsu: Geophys. Mag. *26*, 231-255 (1955)
5.11 H.H. Lamb: *Climate: Present, Past and Future, Climatic History and the Future*, Vol.2 (Methuen, London 1977) p.227
5.12 H.E. Landsberg, J.M. Albert: Weatherwise *27*, 63-66 (1974)
5.13 J.B. Hoyt: Assoc. Am. Georg. *48*, 118-131 (1958)
5.14 H. Stommel, E. Stommel: Sci. Am. *240*, 176-186 (1979)
5.15 D.V. Hoyt: Nature London *275*, 630-632 (1978)
5.16 J. Imbrie, J.Z.Z. Imbrie: Science *207*, 943-953 (1980)
5.17 J.P. Kennett, R.C. Thunell: Science *196*, 1231-1234 (1977)
5.18 A.J. Gow, T. Williamson: Earth Planet. Sci. Lett. *13*, 210-218 (1971)
5.19 L.G. Thompson: "Microparticles, Ice Sheets and Climate"; Instr. of Polar Studies Rept. 64, Ohio State Univ., RF3416-A1 (1977)
5.20 C.U. Hammer: Nature London *270*, 482-486 (1977)
5.21 R.A. Bryson, B.M. Goodman: Science *207*, 1041-1044 (1980)
5.22 J.B. Pollack, O.B. Toon, C. Sagan, A. Summers, B. Baldwin, W. VanCamp: J. Geophys. Res. *81*, 1071-1083 (1976)
5.23 S. Self, M.R. Rampeno, J.J. Barbeda: Paleogeogr. Paleoclimatol. Paleoecol. (in press, 1981)
5.24 J.R. Bray: Nature London *248*, 42-43 (1974)
5.25 L. Koch: Medd. Groenl. *130*, 1-373 (1945)
5.26 S.H. Schneider, C. Mass: Science *190*, 741-746 (1975)
5.27 A. Robock: J. Atmos. Sci. *35*, 1111-1122 (1978)
5.28 A. Robock: Science *206*, 1402-1404 (1979)
5.29 D. Deirmendjian: Adv. Geophys. *16*, 267-296 (1973)
5.30 O.B. Toon, J.B. Pollack: J. Appl. Meteorol. *15*, 225-246 (1976)
5.31 O.B. Toon, J.B. Pollack: Am. Sci. *68*, 268-278 (1980)
5.32 M.P. McCormick, P. Hamill, T.J. Pepin, W.P. Chu, T. Swissler, L.R. McMaster: Bull. Am. Meteorol. Soc. *60*, 1038-47 (1979)
5.33 D.J. Hofmann, J.M. Rosen: J. Geophys. Res. *82*, 1435-1440 (1977)
5.34 B.W. Gandrud, A.L. Lazrus: Geophys. Res. Lett. *8*, 21-22 (1981)
5.35 A.W. Castleman, Jr., H.R. Munkelivitz, B. Manowitz: Tellus *26*, 222-233 (1974)
5.36 A.J. Dyer, B.B. Hicks: Nature London *208*, 131-133 (1965)
5.37 S.C. Mossop: Nature London *203*, 824-827 (1964)
5.38 P.V. Hobbs, L.F. Radke, M.W. Eltgroth, D.A. Hegg: Science *211*, 816-818 (1981)
5.39 E.C.Y. Inn, J.F. Vedder, E.P. Condon, D. O'Hara: Science *211*, 821-823 (1981)
5.40 B.W. Gandrud, A.L. Lazrus: Science *211*, 826-827 (1981)
5.41 T. Vossler, D.L. Anderson, N.K. Aras, J.M. Phelan, W.H. Zoller: Science *211*, 827-830 (1981)
5.42 K.F. Palmer, D. Williams: Appl. Opt. *14*, 208-219 (1975)
5.43 N.H. Farlow, V.R. Oberbeck, K.G. Snetsinger, G.V. Ferry, G. Polkowski, D.M. Hayes: Science *211*, 832-834 (1981)
5.44 J.B. Pollack, O.B. Toon, B.N. Khare: Icarus *19*, 3-2-389 (1973)
5.45 J.A. Ogren, R.J. Charlson, L.F. Radke, S.K. Domonkos: Science *211*, 834-836 (1981)
5.46 E.M. Patterson: Science *211*, 836-838 (1981)
5.47 R.L. Chuan, D.C. Woods, M.P. McCormick: Science *211*, 830-832 (1981)
5.48 J.B. Pollack, O.B. Toon, A. Summers, B. Baldwin, C. Sagan, W. VanCamp: Nature London *263*, 551-555 (1976)

5.49 P. Chylek, J.A. Coakley, Jr.: Science *183*, 75-77 (1974)
5.50 B.C. Weare, R.L. Temken, F.M. Snell: Science *186*, 827-828 (1974)
5.51 J.B. Pollack, O.B. Toon, D. Wiedman: Geophys. Res. Lett. *8*, 26-28 (1981)
5.52 G. Fiocco, G.W. Grams, A. Mugnai: in *Radiation in the Atmosphere*, ed. by H.J. Bolle (Science Press, Princeton, NJ 1976) p.74
5.53 J.A. Ogren, N.C. Ahlquist, A.D. Clarke, R.J. Charlson: Geophys. Res. Lett. *8*, 9-12 (1981)
5.54 J.E. Hansen, W. Wang, A.A. Lacis: Science *199*, 1065-1068 (1978)
5.55 R.P. Turco, O.B. Toon, J.B. Pollack, R.C. Whitten, I.G. Poppoff, P. Hamill: J. Appl. Meteorol. *19*, 78-89 (1980)
5.56 R.P. Turco, R.C. Whitten, O.B. Toon, J.B. Pollack, P. Hamill: Nature London *283*, 283-286 (1980)

Subject Index

Absorption of sunlight 121,136
Absorption of terrestrial radiation 124
ACE Program 46
Aircraft flight effects 121,144,145
see also Supersonic transport
  high-altitude research 1-3,5,15,19, 23-25,31,33,34,46,52,93,134,141
Aitken nuclei 23,24,27,94
Aluminum oxide 23,31,101,113
Ammonia (NH_3) 87
Ammonium (NH_4) 2,25,29,69,83
Anthropogenic influences 95,113,116, 117,121,123,144,145
Aureole method 48

Balloons 1-3,15,16,19,23-25,30,31,33, 34,39,52,60,62,93

Carbon dioxide (CO_2) 144
Carbon disulfide (CS_2) 4,5,15,22,23, 102,104,115,116
Carbonyl sulfide (OCS) 1,4-6,12,15, 18,19,21,22,71,72,90,102-104,110,113, 115,116,135,145
Charge balance 89,93
Chemical reactions 69,74,95
  heterogeneous 69,72,73,77,84-87,89
  homogeneous 73,74
  oxidation 72,73,86,87,90,93
  rate coefficients 72
Climate 6,23,30,49,93,113,116,121-124, 127,128,131,133,136,142-145
Climatic Impact Assessment Program (CIAP) 39

Clustering 5,8,73-79,81,82,89,110
Clusters, critical 75-80,82,84,107
Coagulation 6,7,10,11,69,74,90,93,94, 96,102,105
  kernels 10,11,94,96,98,102
Composition, aerosol 1-3,23-25,28,29, 69,70,79,86,93,98,111
Condensation 6,7,9,10,24,69,73,74,80, 82,83,90,94,96,97,121
  nuclei 101,102,109,112,144
  nuclei counters 107
Contact angle 81,105
Continuity equation, generalized aerosol (GACE) 74,95
Cooling, northern hemisphere in twentieth century 132
Cores
  aerosol 102
  ice 128,130
  ocean bottom 130
Cosmic dust 23
Cosmic rays as ionization source 102
Cross section
  extinction 139,140
  optical ......
  scattering 138
Cryogenic collection (gases) 5,17-20
Cunningham factor 12

Diffusion
  "eddy" 6,7,12,40,96,99
  molecular 84,85
  numerical 100

Dimethyl sulfide $(CH_3)_2S$ 71
Dipole moment   82
Dust veil index   129,131
Dustsonde   40,43,45,52,62,63

Electricity, atmospheric   93,97
Electron affinity   82
Electron attachment   98
Electron microscope   25
Electron spectroscopy (ESCA)   25
Evaporation   2,6,7,9,69,73,84,85,90
Extinction of sunlight by aerosols   23,48-50,54,55,60,63,136,140

Filters, aerosol collection   1-3,5,16,17,24,46
Finite Difference methods   99
  explicit   100
  implicit   100
Franklin, Benjamin   121,122

Gas chromotography   5,18,19,22
Gas-to-particle conversion   7,27,70,93,94,136
Glacial advances   123,127,131
Glasses   31,136
Grains   24,26,29,30,81
Granular inclusions   23
Ground truth   52,63
Growth, particle   5,18,19,22,69,79,84,85,96

HSO_x radicals   5,8,72,83
Henry's law   86
Hydrogen chloride (HCl)   82,87,89
Hydrogen sulfide (H_2S)   70,71
Hydroxyl (OH)   4,5,72,82,90,103,104

Ice ages   123,127-129
  glacial   144
  "little ice age"   131,132,144
  Wisconsin   129
Ice sheets   129

Impaction   1,2,4,24,34
  cascade   1
  plates   1
  quartz crystal   24,52
  wires   1,2,25
Inversion (radiative transfer equation)   48
Ion attachment   89
Ion detachment   98
Ionization potential   82

Kelvin effect   78,85
Knudsen number   10

Laser   35,48
Lidar   1,3,4,34-39,41-43,45,46,48,52,62,93

Magma   136
Mass spectrometer   25
Maunder minimum   132
Mean free path   12
Meteoritic ablation   8
Meteoritic material   95,101,103,105,107-109,112
Meteors   27,30,31,80,85
Microphysical processes   6-12
Microphysical properties   93,98,102
Mie scattering   6,134
Models
  aerosol   6,9,12,21,22,37,46,72,90,99,100,144
  box   99,100
  Eulerian   99
  Fisher droplet   78
  Lagrangian   99
  radiative-convective   140
  radiative transfer   144
  Thomson droplet   77-79,82
  two-dimensional aerosol   100
Moments, method of   99

Nacreous clouds   112

Nephelometer   52,134

Neutron activation   25,30

Nimbus 7 satellite   49,52,54

Nitrates   25

Nitric acid   80,82

Noctilucent clouds   112,113

Nucleation   4,6-9,69,73-75,90,93,94, 96-98,101,105,106
  barrier   75-79,81
  binary   74,79
  heterogeneous   5,7,8,74,78,90,100, 105,106
  heteromolecular   74,76,78,80,84,90, 100
  homogeneous   7,8,74,76-80,83,90
  ion-induced   7,8,81,105
  ternary   74,80
  theory   75,78,80,82,105
  unary   74,77,78

Nucleus   23,24,27
  condensation *see* condensation nuclei
  wettable   81

Opacity   137,140
  infrared   142

Optical constants *see* Refractive indices

Optical depth   116,121,127,130,131, 133-135,138,140-145

Optical methods
  lidar *see* Lidar
  particle counters   1,3,47
  radar *see* Lidar

Optical phenomena   133

Precursor gases   15,17,21,96,97,102

Purple light   1

Quasi-bienniel oscillation   123,124

Radiation
  absorption *see* Absorption of sunlight
  balance   132
  budget   133,141
  infrared   6,121,123
  scattering *see* Scattering of light
  terrestrial   133
  transport   113,116

Radiative equilibrium   140

Radiative properties   121

Rainout *see* Washout

Recombination, ion   89,98,110

Refraction (optical) and refractive indices   3,48,50,63,134,136,138-140

Reynolds number   11

Rocket exhaust products   23,31,101,113

Satellites   1,3,35,46-48,93,133
  SAGE   3,43,45,46,49-52,54,60-64,134
  SAM II   3,45,46,49-52,54,57-60,62-64, 134,141

Scattering of light by aerosols   23,26, 33,36,37,43,48,50,121,133,134
  albedo   136,138-141,145
  backscattering   143
  forward scattering   134,137,139,143
  phase function   134,138

Sedimentation   6,7,11,12,90,93,94,96, 99,108,110

Silicates   135,136

Size distributions   1,2,8,24,26,28-31, 33,34,46,48,69,74,84,94,98,99,102,109, 111,114,134,136-138,140,141,143
  volcanic ash   139
  zero-order logarithmic (ZOLD)   28,138

Solar activity   132

Soot (from aircraft)   6,101,113,145

Space shuttle   113,115,121,144

Spectroscopic method   21,22

Stokes law   11

Stratospheric clouds   58,59

Sulfate   2,3,15,16,25,30,32,59,69-73, 79,90,93,103,106-108,112,116,128,130, 135-137,140

Sulfur 2,16,17,19,29,70,72,83,89,90, 102,103,110,121,136,144
  anthropogenic sources 71
  biogenic 71
  isotopic ratio 70,71
Sulfur dioxide (SO_2) 1,4-6,8,12,15, 17,22,70-72,82,90,94,103,104,110,121, 135,145
  hydrate 73
Sulfur trioxide (SO_3) 5,72,87
Sulfuric acid
  lidquid droplets 5,7,9,10,23,27,29, 30,70,78,79,81-84,86,87,101,103,107, 111,130,134,135,140,143
  vapor 5,7-10,70,72,74,78,80-82,84, 90,102-105,107,108,113,121,135
Supersaturation 73,74,77,78,81-83,90, 105
Supersonic transports 113,115,144,145

Temperature
  stratospheric 123,124,141-143
  surface and tropospheric 6,23,46, 113,116,124-127,141-143,145

Transmission of light through the atmosphere 123,129,130
Transport 6,7,12,34,40,43,69,85,94
  "eddy" 6,7,12,40,96,99,108
Tropopause 15,29,31,32,54,59,60,63,106
Troposphere-stratosphere exchange 3
Tropospheric folding 3,7

Variations
  meridional 3,4,31,34
  temporal 1,3,31,32
Viscosity, air 12
Volcanic ash 23,28,29,31,33,110,111, 121,130,134,136-138,140
Volcanic eruptions and activity 1,4,6, 16,21,27-34,36-43,45,46,48,60-62,70,71, 80,85,89,102,103,108,110,121-136,138, 140,142-145

Warming, twentieth century 132
Washout 96,107

"Year without a summer" 126

Interactions on Metal Surfaces

Editor: R. Gomer

1975. 112 figures. XI, 310 pages
(Topics in Applied Physics, Volume 4)
ISBN 3-540-07094-X

Contents:
J. R. Smith: Theory of Electronic Properties of Surfaces. – *S. K. Lyo, R. Gomer:* Theory of Chemisorption. – *L. D. Schmidt:* Chemisorption: Aspects of the Experimental Situation. – *D. Menzel:* Desorption Phenomena. – *E. W. Plummer:* Photoemission and Field Emission Spectroscopy. – *E. Bauer:* Low Energy Electron Diffraction (LEED) and Auger Methods. – *M. Boudart:* Concepts in Heterogeneous Catalysis.

I. I. Sobelman, L. A. Vainshtein, E. A. Yukov

Excitation of Atoms and Broadening of Spectral Lines

1981. 34 figures, 40 tables. X, 315 pages
(Springer Series in Chemical Physics, Volume 7)
ISBN 3-540-09890-9

Contents:
Elementary Processes Giving Rise to Spectra. – Theory of Atomic Collisions. – Approximate Methods for Calculating Cross Sections. – Collisions Between Heavy Particles. – Some Problems of Excitation Kinetics. – Tables and Formulas for the Estimation of Effective Cross Sections. – Broadening of Spectral Lines. – References. – List of Symbols. – Subject Index. – Errata for volume 1 of this series.

I. I. Sobelman

Atomic Spectra and Radiative Transitions

1979. 21 figures, 46 tables. XII, 306 pages
(Springer Series in Chemical Physics, Volume 1)
ISBN 3-540-09082-7

Contents:
Elementary Information on Atomic Spectra: The Hydrogen Spectrum. Systematics of the Spectra of Multielectron Atoms. Spectra of Multielectron Atoms. – Theory of Atomic Spectra: Angular Momenta. Systematics of the Levels of Multielectron Atoms. Hyperfine Structure of Spectral Lines. The Atom in an External Electric Field. The Atom in an External Magnetic Field. Radiative Transitions. – References. – List of Symbols. – Subject Index.

Turbulent Reacting Flows

Editors: P. A. Libby, F. A. Williams
With contributions by numerous experts

1980. 38 figures, 3 tables. XI, 243 pages
(Topics in Applied Physics, Volume 44)
ISBN 3-540-10192-6

Contents:
P. A. Libby, F. A. Williams: Fundamental Aspects. – *A. M. Mellor, C. R. Ferguson:* Practical Problems in Turbulent Reacting Flows. – *R. W. Bilger:* Turbulent Flows with Nonpremixed Reactants. – *K. N. C. Bray:* Turbulent Flows with Premixed Reactants. – *E. E. O'Brien:* The Probability Density Function (pdf) Approach to Reacting Turbulent Flows. – *P. A. Libby, F. A. Williams:* Perspective and Research Topics.

Springer-Verlag Berlin Heidelberg New York

Aerosol Microphysics I

Particle Interaction

Editor: W. H. Marlow
With contributions by numerous experts

1980. 35 figures, 1 table. XI, 160 pages
(Topics in Current Physics, Volume 16)
ISBN 3-540-09866-6

Contents:
W. H. Marlow: Introduction: The Domains of Aerosol Physics. – *J. R. Brock:* The Kinetics of Ultrafine Particles. – *J. D. Doll:* Classical and Statistical Theories of Gas-Surface Energy Transfer. – *P. J. McNulty, H. W. Chew, M. Kerker:* Inelastic Light Scattering. – *W. H. Marlow:* Survey of Aerosol Interaction Forces.

Inverse Scattering Problems

in Optics

Editor: H. P. Baltes
With contributions by numerous experts
With a Foreword by R. Jost

1980. 49 figures, 2 tables. XIV, 313 pages
(Topics in Current Physics, Volume 20)
ISBN 3-540-10104-7

Contents:
H. P. Baltes: Progress in Inverse Optical Problems. – *G. Ross, M. A. Fiddy, M. Nieto-Vesperinas:* The Inverse Scattering Problem in Structural Determinations. – *E. Jakeman, P. N. Pusey:* Photon-Counting Statistics of Optical Scintillation. – *A. Selloni:* Microscopic Models of Photodetection. – *M. Bertero, C. De Mol, G. A. Viano:* The Stability of Inverse Problems. – *R. Goulard, P. J. Emmerman:* Combustion Diagnostics by Multiangular Absorption. – *W.-M. Boerner:* Polarization Utilization in Electromagnetic Inverse Scattering.

Inverse Source Problems

in Optics

Editor: H. P. Baltes
With contributions by numerous experts
With a foreword by J.-F. Moser

1978. 32 figures. XI, 204 pages
(Topics in Current Physics, Volume 9)
ISBN 3-540-09021-5

Contents:
H. P. Baltes: Introduction. – *H. A. Ferwerda:* The Phase Reconstruction Problem for Wave Amplitudes and Coherence Functions. – *B. J. Hoenders:* The Uniqueness of Inverse Problems. – *H. G. Schmidt-Weinmar:* Spatial Resolution of Subwavelength Sources from Optical Far-Zone Data. – *H. P. Baltes, J. Geist, A. Walther:* Radiometry and Coherence. – *A. Zardecki:* Statistical Features of Phase Screens from Scattering Data.

Raman Spectroscopy

of Gases and Liquids

Editor: A. Weber
With contributions by numerous experts

1979. 103 figures, 25 tables. XI, 318 pages
(Topics in Current Physics, Volume 11)
ISBN 3-540-09036-3

Contents:
A. Weber: Introduction. – *S. Brodersen:* High-Resolution Rotation-Vibrational Raman Spectroscopy. – *A. Weber:* High-Resolution Rotational Raman Spectra of Gases. – *H. W. Schrötter, H. W. Klöckner:* Raman Scattering Cross Sections in Gases and Liquids. – *R. P. Srivastava, H. R. Zaidi:* Intermolecular Forces Revealed by Raman Scattering. – *D. L. Rousseau, J. M. Friedman, P. F. Williams:* The Resonance Raman Effect. – *J. W. Nibler, G. V. Knighten:* Coherent Anti-Stokes Raman Spectroscopy.

Springer-Verlag Berlin Heidelberg New York